Efficient Nutrient Management in California Vegetable Production

Timothy K. Hartz

Emeritus Extension Specialist
Department of Plant Sciences
University of California, Davis

UNIVERSITY OF CALIFORNIA
Agriculture and Natural Resources

Davis, California

Publication 3555
2020

To order or obtain ANR publications and other products, visit the ANR Communication Services online catalog at https://anrcatalog.ucanr.edu/ or phone 1-800-994-8849. You can also place orders by mail or request a printed catalog of our products from

UC Agriculture and Natural Resources
Publishing
2801 Second Street
Davis, CA 95618
Telephone 1-800-994-8849
Email: anrcatalog@ucanr.edu

Publication 3555
ISBN-13: 978-1-62711-070-9

Library of Congress Cataloging-in-Publication Data
Names: Hartz, Tim, author.
Title: Efficient nutrient management in California vegetable production / Timothy K. Hartz.
Other titles: Publication (University of California (System). Division of Agriculture and Natural Resources); 3555.
Description: Oakland, California: University of California Agriculture and Natural Resources, 2020. | Series: Publication / University of California Agriculture and Natural Resources 3555 | Includes bibliographical references and index. | Summary: "This peer-reviewed book outlines principles of nutrient management for commercial vegetable production in California across crops and production regions. It provides growers, consultants, and fertilizer industry personnel with a practical understanding of crop nutrient requirements, soil nutrient availability, the value and limitations of soil and plant nutrient monitoring, and environmental protection."-- Provided by publisher.
Identifiers: LCCN 2019044397 | ISBN 9781627110709 (paperback)
Subjects: LCSH: Vegetables--Nutrition--California. | Vegetables--Fertilizers--California.
Classification: LCC SB321.5.C2 H37 2020 | DDC 635.09794--dc23 LC record available at https://lccn.loc.gov/2019044397

Design by Sandra J. Osterman and Celeste A. Rusconi. Credits are given in the captions.

The University of California, Division of Agriculture and Natural Resources (UC ANR) prohibits discrimination against or harassment of any person in any of its programs or activities on the basis of race, color, national origin, religion, sex, gender, gender expression, gender identity, pregnancy (which includes pregnancy, childbirth, and medical conditions related to pregnancy or childbirth), physical or mental disability, medical condition (cancer-related or genetic characteristics), genetic information (including family medical history), ancestry, marital status, age, sexual orientation, citizenship, status as a protected veteran or service in the uniformed services (as defined by the Uniformed Services Employment and Reemployment Rights Act of 1994 [USERRA]), as well as state military and naval service.

UC ANR policy prohibits retaliation against any employee or person in any of its programs or activities for bringing a complaint of discrimination or harassment. UC ANR policy also prohibits retaliation against a person who assists someone with a complaint of discrimination or harassment, or participates in any manner in an investigation or resolution of a complaint of discrimination or harassment. Retaliation includes threats, intimidation, reprisals, and/or adverse actions related to any of its programs or activities.

UC ANR is an Equal Opportunity/Affirmative Action Employer. All qualified applicants will receive consideration for employment and/or participation in any of its programs or activities without regard to race, color, religion, sex, national origin, disability, age or protected veteran status.

University policy is intended to be consistent with the provisions of applicable State and Federal laws.

Inquiries regarding the University's equal employment opportunity policies may be directed to: Affirmative Action Contact and Title IX Officer, University of California, Agriculture and Natural Resources, 2801 Second Street, Davis, CA 95618, (530) 750-1397. Email: titleix-discrimination@ucanr.edu. Website: https://ucanr.edu/sites/anrstaff/Diversity/Affirmative_Action/.

To simplify information, trade names of products have been used. No endorsement of named or illustrated products is intended, nor is criticism implied of similar products that are not mentioned or illustrated.

This publication has been anonymously peer reviewed for technical accuracy by University of California scientists and other qualified professionals. This review process was managed by ANR Associate Editor for Pest Management—Agricultural, Brenna Aegerter.

TBD-8/19-SB/LC/SO

CONTENTS

PREFACE

Nutrient management is critical to successful vegetable production. Given the high value of vegetable crops and exacting market standards for size, appearance, and postharvest quality, fertilizer management practices have evolved largely to optimize production across a wide range of field conditions. While these practices have been effective in producing good crops, they are not necessarily efficient from the standpoint either of input costs or environmental protection. With growers now facing increasingly stringent regulations designed to minimize nutrient losses to the environment, a general reevaluation of established fertilization practices is warranted.

Commercial vegetable production in California is a complex undertaking. Dozens of vegetable crops are produced on a year-round basis in production areas with very different environmental conditions. Given the complexity of the industry, it is impractical to produce a detailed manual outlining fertility management practices specific to each crop and production region. The intent of this publication is to outline the principles of nutrient management that are broadly applicable across crops and production regions. Unlike existing reference materials on soil fertility and nutrient management, this publication is specific to California vegetable crop production; the principles discussed are illustrated using examples drawn from recent California research. This publication provides growers, consultants, and fertilizer industry personnel with a practical understanding of crop nutrient requirements, soil nutrient availability, the value and limitations of soil and plant nutrient monitoring, and environmental protection.

It has been my privilege to work with vegetable growers and other industry professionals across California on nutrient management issues for the past 25 years. Those interactions have generated much of the information presented here.

ACKNOWLEDGMENTS

This publication is a synthesis of information developed through many years of field research and demonstration projects. The California Department of Food and Agriculture Fertilizer Research and Education Program, the California Leafy Greens Research Board, and the California Tomato Research Institute provided the majority of the funding for the projects, and that funding was augmented by generous donations and in-kind contributions from vegetable growers, fertilizer suppliers, and other industry groups across California. Many University of California colleagues assisted in the development of the information presented here. I particularly want to acknowledge the contributions of University of California Cooperative Extension farm advisors Richard Smith, Mike Cahn, and Gene Miyao, as well as former graduate students Paul Johnstone, Tom Bottoms, and Sebastian Castro Bustamante. These groups and individuals made this publication possible.

Soil Nutrients and Nutrient Availability

ESSENTIAL ELEMENTS IN SOIL

Eighteen elements have been identified as essential for plant growth and development. Carbon (C), oxygen (O), and hydrogen (H) are obtained from air and water and comprise the majority of plant biomass; cellulose, starch, and sugars are composed primarily of these elements. Plants obtain the other fifteen essential elements from soil (table 1.1). Plants take up nutrients from the soil either as positively charged ions (cations) or negatively charged ions (anions). The exception is B, which is mostly taken up as undissociated boric acid (H_3BO_3), although in alkaline soil some anionic forms may also be used.

Macronutrients are elements required in large quantities by plants. This group can be divided into the primary macronutrients N, P, and K and the secondary macronutrients Ca, Mg, and S. Primary macronutrients must often be applied in fertilizers or soil amendments to achieve peak crop production and quality. Secondary macronutrients are usually

TABLE 1.1.

Essential elements required for plant growth and their relative abundance in plants

| Element | Symbol | Form(s) taken up from soil | | Typical leaf concentration range (dry weight) | Typical crop nutrient content at harvest (elemental lb/ac)* |
		Formula	Name		
nitrogen	N	NO_3^-, NH_4^+	nitrate, ammonium	2–5%	270
phosphorus	P	HPO_4^{2-}, $H_2PO_4^-$	hydrogen phosphate, dihydrogen phosphate	0.2–0.6%	40
potassium	K	K^+	potassium	2–5%	350
calcium	Ca	Ca^{2+}	calcium	0.5–4.0%	200
magnesium	Mg	Mg^{2+}	magnesium	0.4–1.0%	80
sulfur	S	SO_4^{2-}	sulfate	0.25–1.0%	80
chlorine	Cl	Cl^-	chloride	0.3–1.5%	60
iron	Fe	Fe^{2+}, Fe^{3+}	iron	50–250 ppm	1
manganese	Mn	Mn^{2+}	manganese	30–250 ppm	0.4
boron	B	H_3BO_3	boric acid	20–80 ppm	0.3
zinc	Zn	Zn^{2+}	zinc	20–70 ppm	0.3
copper	Cu	Cu_2^+	copper	5–20 ppm	0.1
molybdenum	Mo	MoO_4^{2-}	molybdate	< 2 ppm	0.01
nickel	Ni	Ni^{2+}	nickel	< 1 ppm	< 0.01
cobalt	Co	Co^{2+}	cobalt	< 1 ppm	< 0.01

*Based on a typical processing tomato crop; values given represent crop *uptake*, which for some elements may be substantially higher than the actual crop *requirement*.

sufficiently abundant in soil to maximize crop growth without additional input, but Ca and S may be applied to soil for purposes unrelated to plant nutrition. Macronutrient concentrations in plant tissue are high enough that they are typically expressed as a percentage of dry matter.

Micronutrients are elements required in small quantities by the plant. They include Fe, Mn, Zn, Cu, B, Cl, Mo, Co, and Ni. All except Cl are present in plant tissue at a low concentration in the parts per million (ppm) range. While only a small quantity of Cl is required by plants, relatively high levels are often present in soil, and consequently plant uptake can be relatively high (more than 1 percent of plant dry weight in some cases). California soils are generally well supplied with micronutrients, although their distribution and plant availability vary regionally and among fields within regions. Micronutrient fertilization is required in only a small percentage of vegetable fields. Cobalt and nickel are required in minute quantity by plants, and there is no record of soil deficiency of either element in California vegetable production; these elements are of no agronomic significance and do not require active management.

SOIL PROPERTIES AFFECTING NUTRIENT AVAILABILITY

Certain fundamental properties of soil govern to a large extent the availability and retention of nutrients. A general understanding of those properties is essential to efficient soil fertility management.

Soil pH

Soil pH is a measure of its acidity or alkalinity as indicated by the negative logarithm of the hydrogen (H^+) ion concentration on a scale from 0 to 14. A pH of 6.0 indicates a 10-fold higher H^+ concentration than a pH of 7.0. Soils with pH less than 7.0 are considered acidic, while soils with pH greater than 7.0 are considered alkaline. California soil is typically in the 5.5 to 8.0 range.

Soil pH has a profound influence on soil chemistry and nutrient availability. As pH gets progressively more acidic, H^+ and aluminum (Al^{3+}) ions displace Ca^{2+}, Mg^{2+}, and K^+ ions from cation exchange sites, making them subject to leaching. Soil pH also affects the availability of phosphorus and most micronutrients. As pH gets progressively more alkaline, these nutrients become less available to plants.

Cation and Anion Exchange

Soil particles have many negative charges on their surfaces that attract cations. The amount of cations that can be adsorbed to these negative charges is referred to as the cation exchange capacity (CEC). Soil particles also contain positively charged sites; the ability of soil to adsorb anions on these sites is referred to as the anion exchange capacity (AEC). However, soil AEC is very limited, particularly in alkaline soil, and is of little agronomic significance. Therefore, anions move more freely through soil with rainfall or irrigation, while cations are more resistant to leaching.

The dominant cations in soil with pH greater than 7.0 are Ca^{2+}, Mg^{2+}, K^+, and sodium (Na^+), generally in that order of abundance in most California soils. They are collectively referred to as base cations. These cations are held with different strengths on the cation exchange. From most strongly held to most weakly held, the order is $Ca^{2+} > Mg^{2+} > K^+ > Na^+$. Consequently, K^+ and Na^+ are somewhat more mobile in soil than are Ca^{2+} or Mg^{2+}. However, the leaching potential of K^+ and Na^+ is still much less than for anions (NO_3^-, SO_4^{2-}, etc.).

In acid soils two other cations play an indirect role in soil fertility through their effect on the cation exchange: Al^{3+} and H^+. Both are strongly held on the cation exchange, so when they are present in substantial quantities they can displace the base cations and make them subject to leaching. Aluminum is a constituent of common soil minerals and is present in all soils. However, the solubility of Al compounds increases in acid soils, so its share of the CEC depends on the soil pH. Above pH 5.5, the amount of Al^{3+} on the cation exchange is relatively insignificant, but below that pH Al^{3+} increasingly displaces the base cations. Similarly, the fraction of CEC occupied by H^+ is pH dependent, ranging from essentially none above pH 7.0 to about 30 percent at pH 5.5.

Base saturation captures the relative abundance of the base cations on the cation exchange. It is calculated as the fraction of the cation exchange occupied by Ca^{2+}, Mg^{2+}, K^+, and Na^+. The lower the percentage of base saturation, the lower the soil's capacity to supply Ca, Mg, and K.

Texture and Structure

Soil contains mineral particles of different sizes. Particle size distribution has a tremendous direct influence on soil fertility and an indirect influence through its effect on soil water-holding capacity. Three particle size categories are recognized: sand (0.05 to 2.0 mm), silt (0.002 to 0.05 mm) and clay (less than 0.002 mm). To put these size ranges in perspective, the largest sand particles are roughly the thickness of a nickel, while individual clay particles are invisible to the naked eye (fig. 1.1).

Large particles are less chemically active than smaller particles due to a smaller ratio of surface area to volume. Therefore, silt and

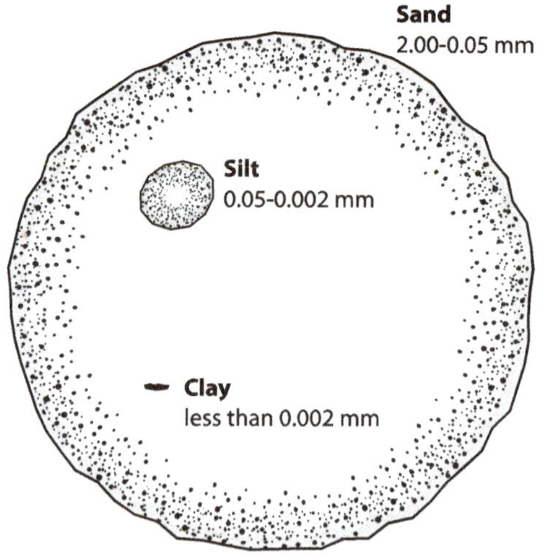

FIGURE 1.1

Relative size of sand, silt, and clay soil particles. Source: Science Learning Hub—Pokapū Akoranga Pūtaiao, University of Waikato, sciencelearn.org.nz.

clay particles are a dominant influence on soil properties like CEC; in general, the higher the silt and clay content, the higher the CEC. Twelve soil texture classes are recognized, defined by the percentage of soil dry weight represented by sand, silt, and clay particles (fig. 1.2). The dominant influence of smaller soil particles is demonstrated by the fact that a soil with equal amounts of sand, silt, and clay is called a clay loam.

Soil texture affects fertility through its influence on CEC; texture also affects the leaching potential of anionic nutrients through its effect on soil water-holding capacity and water infiltration rate. A sandy soil has relatively large pore spaces (macropores); soils dominated by silt and clay particles have more overall pore space, but it is mostly comprised of small pores (micropores, fig. 1.3). Therefore, a fine-textured soil holds more water and drains more slowly than a sandy soil.

The term *structure* refers to the way soil particles are arranged. When soil particles clump together to form aggregates, more macropores are created, increasing the water infiltration rate and improving aeration. In the absence of aggregation, soils may drain poorly and become compacted. Soil structure can influence soil nutrient supply in several important ways. Good soil structure promotes high rooting density, which maximizes the

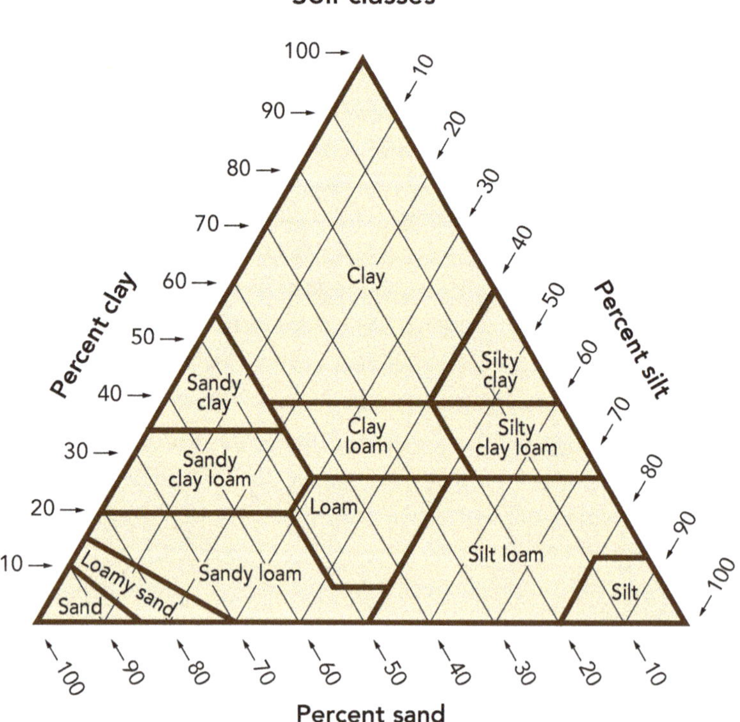

FIGURE 1.2

Soil textural class based on percentage of sand, silt, and clay.

Coarse-textured soil

More macropores, but less
total pore space

Fine-textured soil

More total pore space, but
fewer macropores

FIGURE 1.3

Effect of soil texture on soil porosity.

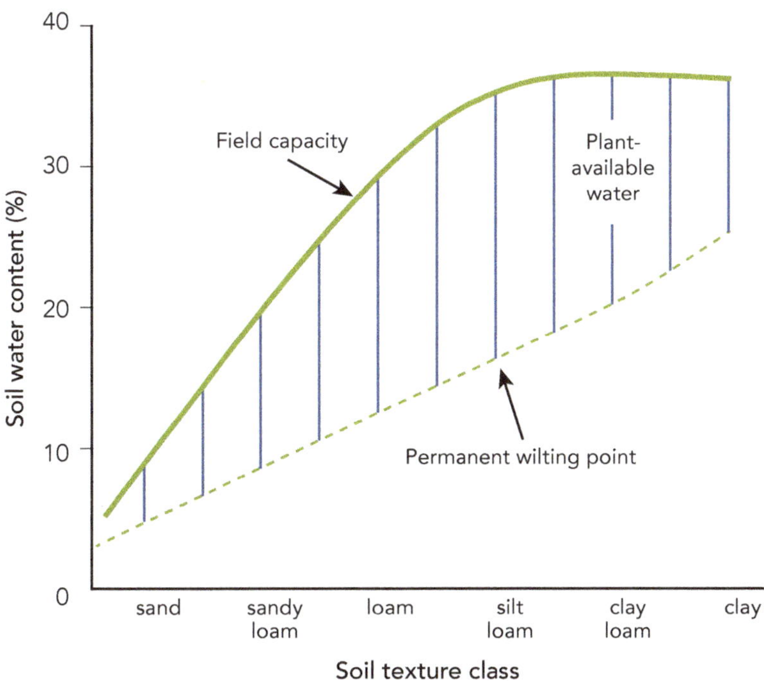

FIGURE 1.4

Effect of soil texture on soil water-holding capacity.

availability of nutrients (mainly K and P) that come into contact with roots primarily by diffusive movement. Texture also affects the water infiltration rate and the degree of aeration in the root zone, both of which affect a plant's ability to access soil nutrients.

Water-Holding Capacity

Soil texture and structure are primary factors controlling how much water a soil can hold and what fraction of that water is plant available. Pore space makes up about half the volume of a typical soil. The fraction of that pore space from which water will drain under the pull of gravity determines a soil's water-holding capacity and its degree of aeration. Sandy soil has larger pores than does more finely textured soil; consequently, once drainage is complete following rainfall or irrigation, sandy soil retains less water and more pore space is filled with air. The amount of water a soil can hold against gravity is referred to as its field capacity. However, only a portion of that water is accessible to plants. As plants remove water from soil, the remaining water is more tightly bound to soil particles by capillary forces; the point at which plants can extract no more water from soil is called the permanent wilting point. The amount of plant-available soil water is the difference between the field capacity moisture content and the permanent wilting point.

The influence of soil texture on soil water-holding properties is shown in figure 1.4. The available water-holding capacity of a sandy soil can be as low as 5 percent by volume, or approximately 0.6 inches of available water per foot of soil. By contrast, water-holding capacity of a silt loam may be nearly 20 percent by volume, equivalent to more than 2 inches of available water per foot of soil. Although texture is the primary factor controlling soil water-holding capacity, soil organic matter content can also have a measurable effect. This is because the water-holding capacity of organic matter, on a percentage of weight basis, is much larger than that of mineral soil particles.

Soil with low water-holding capacity is difficult to irrigate efficiently and is prone to greater fluctuation in the concentrations of anionic nutrients (NO_3^-, SO_4^{2-}) because even a small amount of excess irrigation can leach these ions below the root zone. Consequently, sandy soil may require more frequent N application, and high N use efficiency is harder to achieve.

Organic Matter

Soil organic matter can influence soil fertility in a variety of ways. It can be a source of nutrients, increase cation exchange, and increase both water-holding capacity and water infiltration rate. Soil organic matter is a complex mix of materials; as a generalization one can think of soil organic matter as being composed of active and resistant fractions. The active fraction is made up of soil microorganisms and relatively recently incorporated crop residues or organic amendments. The resistant fraction is humus, older material that is highly resistant to further breakdown. The distinction between the active and resistant fractions is important. The majority of plant-available nutrients released from soil organic matter comes from the active fraction. The resistant fraction of soil organic matter can substantially influence soil structure, water-holding, and CEC, but its effect on soil fertility is limited.

The content of soil organic matter in California soils is generally low, with values of 0.5 to 3.0 percent of soil dry weight being typical. Such soils are referred to as mineral soils, in that their characteristics are dominated by the mineral (nonorganic) fraction. However, soils with much higher levels of soil organic matter (over 20 percent in some cases) are common in a few areas of the state, primarily the Sacramento–San Joaquin Delta and Tulelake regions; such soils are referred to as muck or peat soils.

MECHANISMS OF NUTRIENT UPTAKE

The water contained in soil is referred to as the soil solution. It is from soil solution that the vast majority of nutrient uptake occurs. For most nutrients there is an equilibrium between the nutrient concentration in the soil solution and the abundance of that nutrient elsewhere in the soil: as plants remove a nutrient ion from the soil solution, the equilibrium reaction replaces it from these other sources. For example, Ca ions on cation exchange sites are in equilibrium with Ca ions in soil solution; plant uptake of Ca from the soil solution

TABLE 1.2.

Mechanisms of nutrient uptake and their relative significance for individual nutrients

Element	Ion type	Significance of uptake mechanism		
		Root interception	*Mass flow*	*Diffusion*
nitrogen	anion		***	
phosphorus	anion			***
potassium	cation		*	**
calcium	cation	*	**	
magnesium	cation	*	**	
sulfur	anion		***	
iron	cation		*	**
manganese	cation	*	**	
boron	undissociated		***	
zinc	cation	*	*	*
copper	cation		***	

Source: Adapted from Havlin et al. 2014.

*, **, and *** indicate that uptake mechanism is of slight, moderate, or great significance.

results in movement of Ca from exchange sites into the soil solution. Nitrogen is an exception, in that there is no ready reserve of plant-available N to replenish the soil solution as crop uptake occurs. Nitrogen in soil solution must be maintained through addition of fertilizer N or mineralization of organic N through microbial action.

Nutrients come in contact with the surface of roots, which is the necessary precursor to plant uptake, in three ways:

- root interception, in which a root physically encounters a nutrient ion as it grows

- mass flow, which is the movement of nutrient ions in soil solution to a root surface due to plant transpiration

- diffusion, in which ions move from areas of higher concentration to areas of lower concentration: as a plant root takes up a nutrient from the surrounding soil solution, the concentration of that nutrient near the root is reduced, creating a concentration gradient resulting in the diffusive movement of that nutrient toward the root

Table 1.2 ranks the relative importance of these uptake mechanisms for specific nutrients. Root interception is a very limited phenomenon, accounting for a small fraction of uptake for a few cationic nutrients. Mass flow accounts for the overwhelming majority of N uptake, in the form of nitrate (NO_3^-), and a large fraction of the uptake of most other nutrients as well. Diffusion is responsible for the majority of P and K uptake. Diffusion is so important in P and K availability because their typical concentrations in the soil solution are relatively low compared with the high plant requirement for these elements.

Nutrient uptake by plant roots is not a passive process. Plants have vastly different requirements for the different nutrients (see table 1.1) and therefore must be able to selectively take up the ions they need and exclude those they do not. This can be illustrated by the fact that plants take up much more K than Mg, while most soils contain much more plant-available Mg than K. Furthermore, plants must be able to take up required nutrients against a concentration gradient. For example, NO_3^- and K^+ concentrations are usually much higher in root tissue than in soil solution. This selective uptake of nutrients, often against a concentration gradient, is accomplished by a complex system of transport proteins and biochemical reactions; for an in-depth discussion of nutrient uptake mechanisms, see Marschner (2012). Although the protein transport systems are nutrient specific, they are not perfectly efficient; a serious soil imbalance between chemically similar ions (e.g., K^+ and Mg^{2+}) can result in reduced uptake of a needed nutrient and in some cases the accumulation of a detrimental one. Soil nutrient ratios are important to consider in some cases, but, in general, nutrient ratios must be seriously unbalanced before they have significant agronomic effects.

REFERENCES

Havlin, J. L., S. L. Tisdale, W. L. Nelson, and J. D. Beaton. 2014. Soil fertility and fertilizers. Upper Saddle River, NJ: Pearson.

Marschner, P. 2012. Mineral nutrition of higher plants. London, UK: Academic Press.

2

Crop Growth and Nutrient Requirements

PLANT GROWTH AND NUTRIENT UPTAKE

Annual vegetable crops show a characteristic pattern of growth (fig. 2.1). Whether they are started from seed or transplants, growth in the initial weeks after planting is slow; only about 10 percent of final crop biomass is typically developed in the first third of the growing season. Following the establishment period, growth accelerates in a prolonged rapid growth phase. Vegetative crops such as lettuce and celery are harvested at the end of this rapid growth phase, but crops such as potato and processing tomato go through maturation and senescence, during which growth slows. The amount of seasonal biomass developed varies widely among crops, from about 2,000 pounds per acre of dry biomass in baby spinach to more than 14,000 pounds per acre in processing tomato. Much of the variation among crops relates to the difference in the length of the growing season: spinach may be harvested in as little as 30 days from seeding, while processing tomato may be harvested in 120 days from transplanting.

Nutrient uptake follows this same pattern (fig. 2.2). The nutrient uptake depicted in figure 2.2 is for processing tomato; although the total amount of nutrients taken up varies widely among crops, the seasonal pattern of uptake is quite similar. During the rapid uptake phase, most vegetable crops typically take up from 3 to 7 pounds of N per acre per day. Crops grown during warm conditions and those sown at high density (e.g., spinach) tend to have peak uptake rates on the higher end of that range, while those grown under cooler conditions are at the lower end. Across crops, there are characteristic ratios of seasonal N, P, and K uptake: the N:K ratio typically ranges from 1:1.5 to 1:1, while the N:P ratio is usually from 7:1 to 10:1. These ratios are based on elemental P and K, not P_2O_5 or K_2O equivalents.

The amount of macronutrients taken up depends partly on the vigor of the crop (how much biomass is produced) and also on the level of soil nutrient availability. If the soil nutrient supply is high, crops take up nutrients in excess of what is required to support optimal growth and product quality; this is termed luxury consumption. Fertilizer management that provides excessive soil nutrient availability promotes luxury consumption, wasting money and increasing the potential for nutrient loss to the environment.

Vegetable crops differ widely in the percentage of crop nutrient uptake that is

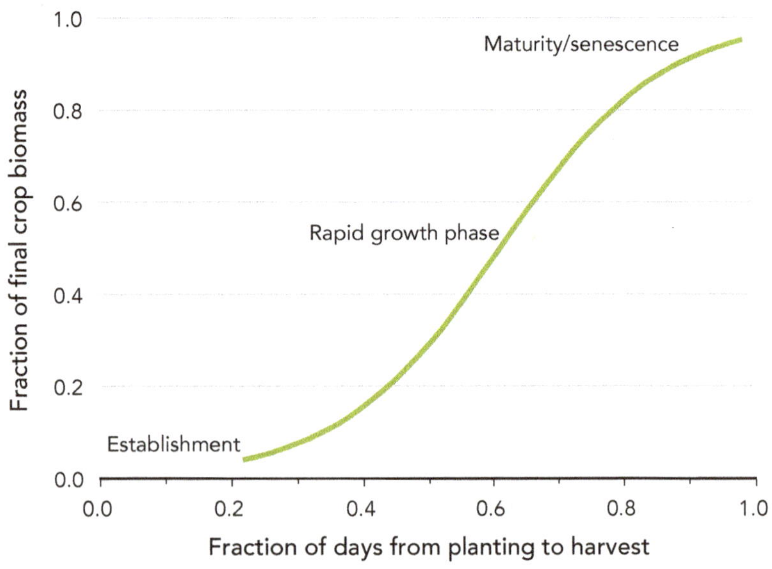

FIGURE 2.1

Typical pattern of growth of annual vegetable crops.

TABLE 2.1.

Typical macronutrient uptake and harvest removal of annual vegetable crops at normal yield levels

Crop	Seasonal crop uptake (lb/ac)*			% nutrient removed with harvest
	N	P	K	
broccoli	250–350	40–50	280–380	25–35
Brussels sprouts	350–500	40–60	300–500	30–50
cabbage	280–380	40–50	300–400	50–60
cantaloupe	150–200	15–25	170–250	50–65
carrot	150–220	25–40	200–300	55–65
cauliflower	250–300	40–45	250–300	25–35
celery	200–300	40–60	300–500	50–65
head or romaine lettuce	120–160	12–16	150–200	50–60
baby lettuce	60–70	5–7	80–100	65–75
onion	150–180	25–35	200–260	60–75
pepper (bell)	240–350	25–50	300–450	55–65
potato	170–250	30–40	250–300	65–75
processing tomato	220–320	35–45	300–400	55–65
spinach	90–130	12–18	150–200	65–75

Sources: Breschini and Hartz 2002; Bottoms et al. 2012; Hartz et al. 2000; Smith et al. 2016a and 2016b; Hartz unpublished data.

*Elemental macronutrient content of aboveground biomass at harvest plus harvested roots (carrot) or tubers (potatoes). To convert table values to P_2O_5 or K_2O, multiply by 2.3 and 1.2, respectively.

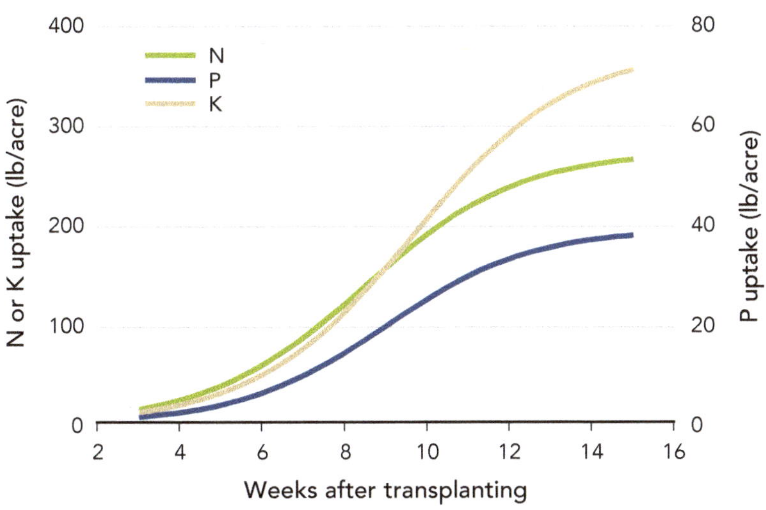

FIGURE 2.2

Pattern of macronutrient uptake by processing tomato.
Source: Adapted from Hartz 2009.

removed from the field with the harvest. This can range from more than 70 percent of macronutrients removed with the harvest of baby greens (lettuce, spinach, etc.) to less than 30 percent removed for broccoli and cauliflower. Most crops are in the 50 to 70 percent range. Nutrient removal in harvest is important for two reasons. Long-term soil fertility maintenance requires replacement of nutrients removed with harvested products; nutrients in crop residue are recycled upon soil incorporation. Also, regulations being instituted in California to minimize nitrate pollution of groundwater will evaluate N efficiency on the basis of the N balance (N applied versus harvest N removal; see chapter 12).

Table 2.1 lists the typical amount of macronutrient uptake by various vegetable crops and the percentage of the total crop macronutrient content removed at harvest. Nutrient uptake by crops grown under cool conditions is generally in the lower end of these ranges, and those grown in the summer are at the higher end. Individual fields can vary substantially from these norms, depending on soil nutrient availability, crop vigor, and marketable yield.

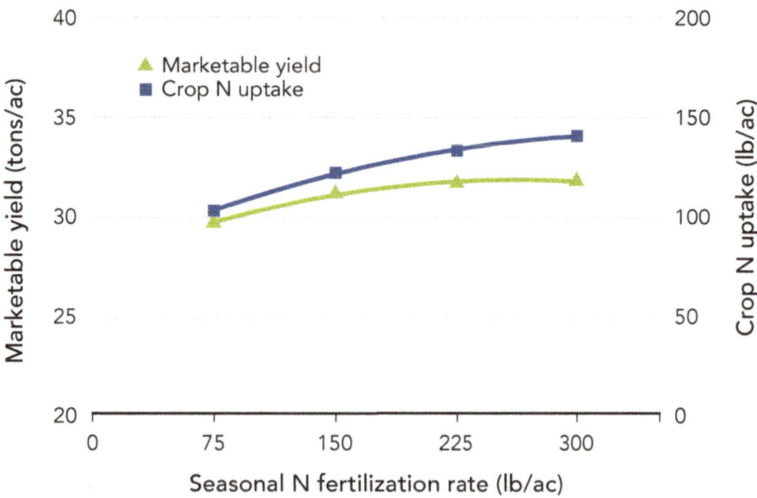

FIGURE 2.3

Effect of seasonal N fertilization rate on lettuce N uptake.
Source: Adapted from Bottoms et al. 2012.

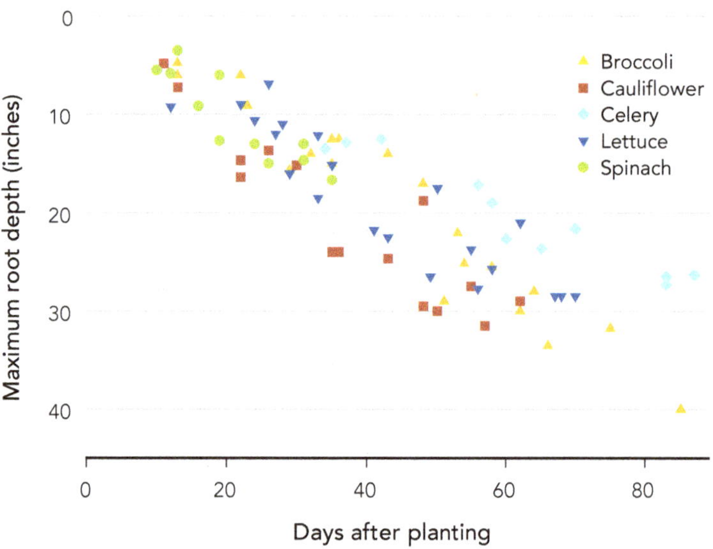

FIGURE 2.4

Pattern of root development in cool-season vegetable crops under summer coastal conditions. Data represent the deepest root observed on each evaluation date.
Source: Smith et al. 2016b and unpublished Salinas Valley studies.

The values in table 2.1 are based on the nutrient uptake observed in commercial fields; because vegetable crops are usually heavily fertilized, these values undoubtedly include some luxury consumption. Some amount of luxury consumption is unavoidable, particularly for K. Crops grown in fields with naturally high soil K availability may take up 20 to 30 percent more K than required for a given yield level. However, luxury consumption of N and P usually reflects grower fertility practices and is therefore largely under the grower's control. It is important to understand that there is a limit to luxury consumption. As soil nutrient availability increases above the level necessary to adequately supply crop requirements, the efficiency with which the crop takes up that nutrient declines. Figure 2.3 illustrates this phenomenon. In this field trial lettuce yield was statistically maximized by applying 150 pounds of N per acre, and total crop N uptake at that level of fertilization was 121 pounds per acre. Applying another 150 pounds per acre of fertilizer N did not significantly affect yield and increased crop N uptake by only 15 pounds per acre. In other words, the N uptake efficiency of that extra 150 pounds of N per acre was only 10 percent, meaning that 90 percent of that fertilizer N remained in the soil and at risk of loss to the environment.

ROOTING DEPTH AND DENSITY

Soil nutrient availability is strongly affected by the depth and density of rooting. Recent research in coastal leafy greens fields has shown that root development continues throughout the growing season (fig. 2.4). While the maximum depth of rooting varied among crops, that difference was largely a function of length of the growing season. The rate of root expansion was similar among crops normally thought of as shallow rooted (lettuce and spinach) and those categorized as deep rooted (broccoli and cauliflower). Surprisingly, whether the crop was seeded or transplanted had little effect on rooting depth.

Even crops that achieve a rooting depth of 3 feet or more concentrate their roots in the top half of the rooting zone (fig. 2.5). This fact, coupled with the reality that soil macronutrient availability tends to decline with increasing soil depth, suggests that effective nutrient management should generally focus on the top foot of soil. An exception to this rule may be the consideration of residual soil NO_3-N below the top foot (see chapter 9 for more detail).

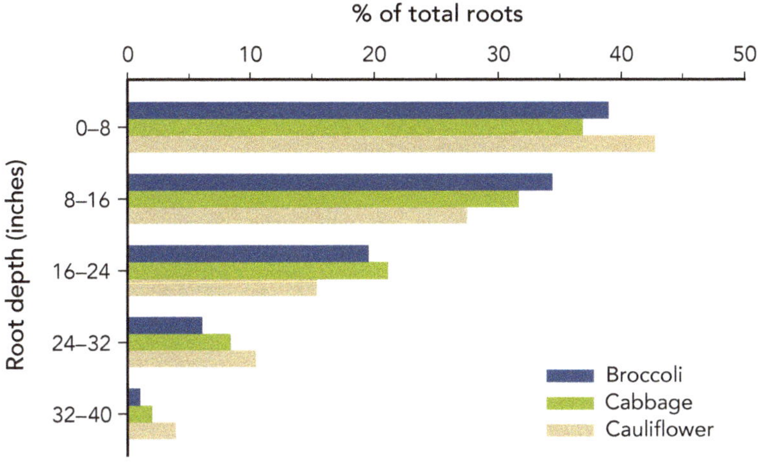

FIGURE 2.5

Root distribution at harvest as a function of soil depth.
Source: Smith et al. 2016b.

TABLE 2.2.

Relative mobility of essential elements in plants

Element	Symbol	Relative mobility
nitrogen	N	high
phosphorus	P	high
potassium	K	high
calcium	Ca	low
magnesium	Mg	high
sulfur	S	high
chlorine	Cl	high
iron	Fe	intermediate
manganese	Mn	intermediate
boron	B	variable*
zinc	Zn	intermediate
copper	Cu	intermediate
molybdenum	Mo	intermediate

Source: Adapted from Epstein and Bloom 2005.

*Dependent on crop species.

The Law of Limiting Factors

There is a widespread perception that when a crop is not exhibiting rapid growth it can be "pushed" by applying additional fertilizer; the nutrient of choice for this purpose is usually N. A particular field may lack vigor for many reasons, but given the generally ample N fertilizer rates used in California vegetable production, lack of N availability is seldom the cause. The law of limiting factors states that plant growth is limited by whatever required factor is in shortest supply. That factor may be a nutrient, or more commonly it may be an environmental variable (temperature, light, soil moisture, etc.) or a pest issue. Applying additional N fertilizer to a field already adequately supplied with N does not increase the rate of crop growth and does not overcome the limitation of cold weather, soil compaction, poor drainage, root damage by a soilborne pathogen, and so on.

NUTRIENT MOBILITY

Plants have two types of conductive tissues that distribute water, nutrients, and organic compounds. Xylem conducts water and mineral nutrients upward from the roots to all aboveground plant parts. Phloem distributes products of photosynthesis, other organic compounds, and some mineral nutrients from the leaves to the rest of the plant. The nutrients N and K are highly mobile in all common crop plants and move freely in both the xylem and the phloem systems (table 2.2).

In addition to rooting depth, root density is also an important factor in soil availability of P and K, which rely on diffusion to move toward plant roots (see table 1.2). The distance that P or K can move in soil by diffusion is limited to less than 1 millimeter per day. Therefore, a relatively high density of roots is required for a plant to effectively access these nutrients.

At the other end of the spectrum, Ca is immobile in plants; it readily moves through the xylem in the transpiration stream, but phloem transport is very limited. Other elements fall between these extremes, with some important differences among crop species: P, Mg, and S are reasonably mobile, while the micronutrients are generally less mobile.

Nutrient mobility also provides clues to the diagnosis of nutrient deficiency symptoms in plants. The high mobility of N and K means that deficiency symptoms are initially expressed on older leaves, while deficiency of less-mobile elements (i.e., Zn and Fe) is expressed on younger leaves.

REFERENCES

Bottoms, T. G., R. F. Smith, M. D. Cahn, and T. K. Hartz. 2012. Nitrogen requirements and N status determination of lettuce. HortScience 47:1768–1774.

Breschini, S. J., and T. K. Hartz. 2002. Drip irrigation management affects celery yield and quality. HortScience 37:894–897.

Epstein, E., and A. J. Bloom. 2005. Mineral nutrition of plants: Principles and perspectives. Sunderland, MA: Sinauer Associates.

Hartz, T. K. 2009. Development of practical fertility monitoring tools for drip-irrigated vegetable production. CDFA-FREP project 06-0626 final report. CDFA website, www.cdfa.ca.gov/is/ffldrs/frep/pdfs/completedprojects/06-0626Hartz.pdf

Hartz, T. K., W. E. Bendixen, and L. Wierdsma. 2000. Pre-sidedress soil nitrate testing as a nitrogen management tool in irrigated vegetable production. HortScience 35:651–656.

Smith, R., M. Cahn, and T. Hartz. 2016a. Evaluation of N uptake and water use of leafy greens grown in high-density 80-inch bed plantings and demonstration of best management practices. CDFA-FREP Project 12-0362 final report. CDFA website, www.cdfa.ca.gov/is/ffldrs/frep/pdfs/completedprojects/12-0362-SA_Smith.pdf.

Smith, R., M. Cahn, T. K. Hartz, P. Love, and B. Farrara. 2016b. Nitrogen dynamics of cole crop production: Implications for fertility management and environmental protection. HortScience 51:1586–1591.

Soil and Water Testing

Evaluation of the nutrient status of soil and irrigation water is the foundation of efficient nutrient management. In addition to the measurement of nutrient status, other characteristics of soil and water can affect nutrient availability and crop growth. Correct interpretation of analytical results requires an understanding of the test methods used. This chapter explains the common analytical methods used by commercial testing laboratories and provides diagnostic guidelines.

SOIL SAMPLING

How a soil sample is collected can have a huge impact on the analytical results. Recommendations vary as to the soil depth to sample, but for general soil fertility evaluation one should sample the primary rooting zone. Based on the rooting patterns illustrated in figures 2.4 and 2.5, sampling the top 12 inches of soil would be an appropriate practice, although some authorities recommend sampling only the top 6 inches. Soil P, K, and micronutrient content typically declines with increasing soil depth, so be aware that lab test values for these nutrients may be as much as 20 to 30 percent lower with a 12-inch sample depth than with a 6-inch sample depth. Soil tillage can affect the relationship between depth and nutrient availability; in general, the less aggressive the tillage, the greater the stratification of nutrient availability by depth. Whatever sampling depth you use, be consistent so you can accurately evaluate trends over time. Use a sampling device that collects the same volume of soil over the entire sampling depth. A soil probe or auger collects a uniform sample, while a sample collected by shovel is likely to overrepresent the top few inches of soil.

A related issue is where to sample in fields with semipermanent beds, as with buried drip irrigation. One might expect that over the course of time nutrient availability in the concentrated root zone around the drip tape may change relative to the soil on the bed edges or in the furrow. A recent study of drip-irrigated processing tomato fields found that soil P and K availability remained relatively consistent across the bed and that collecting cores from varying distances up to 20 inches from the drip tape was a reasonable practice (Lazcano et al. 2015).

Care must be taken to collect a sample that encompasses the variability within a field. Soil is never completely uniform across a field, so multiple cores from various parts of the field must be collected. At least fifteen to twenty soil cores should be collected throughout the field and thoroughly mixed to form a representative sample. Although additional precision may be gained by collecting more cores, the improvement is likely to be minimal. Where significant differences exist in soil type or texture, the collection of separate composite samples from each zone is warranted. Grid sampling, in which soil samples are collected spatially across a field and analyzed separately to create a map of soil nutrient availability, can provide useful information. However, grid sampling can be expensive, and the information is of limited utility without the ability to make a variable-rate fertilizer application. A one-time grid sampling may be useful in directing subsequent sampling to areas of a field most likely to be nutrient limited.

The frequency of soil sampling depends on the information desired. Sampling every 2 years is usually adequate for general soil fertility evaluation because P, K, and micronutrient availability tend to change slowly. An exception may be in very sandy soils, which have limited nutrient reserves and a greater potential for nutrient leaching. Additionally, annual sampling is justified where multiple crops per year are produced, organic or mineral amendments (compost, gypsum, etc.) are applied, or pH is adjusted with lime or sulfur. The time of year that samples should be taken is flexible, the primary criterion being that sampling be done in a manner allowing timely fertilizer application.

Nitrate-nitrogen (NO_3-N) is the soil nutrient most likely to change rapidly over time; consequently, the timing of soil NO_3-N sampling is critical. The NO_3-N concentration of a fall soil sample is unreliable to inform N management decisions for the following spring crop. There is no way to know how much of that NO_3-N may be leached from the root zone with winter rain; also, a fall sample does not reflect the N mineralized over the winter from crop residue or fall-applied organic amendments. A preplant soil NO_3-N sample can be useful in N management, provided that irrigation for crop establishment is managed to minimize leaching. Soil NO_3-N sampling is most useful when conducted in season to inform N application decisions (see chapter 9).

Postcollection handling of soil samples has only a marginal effect on most soil constituents. Samples can be delivered to the testing lab either field moist or air dried. An exception is NO_3-N: since the NO_3-N concentration can change as long as soil microbial activity continues, rapid air-drying or quick cooling of the sample before delivery to the lab is recommended. Another possible exception is K. Drying soil may change the amount of K that is extracted with common lab tests, and some midwestern universities are now recommending testing of field-moist soil. However, the vast majority of western labs air-dry samples before analysis, and all California calibration data for soil test K results are based on analysis of dry soil.

IRRIGATION WATER SAMPLING

Sampling well water for mineral constituents requires attention to several issues. The well should be run for a minimum of 30 minutes before sample collection, and the sample should be taken as close to the well as possible to minimize contamination from pipes or equipment. Testing labs may provide sampling bottles, or plastic drinking water bottles may be rinsed and reused. The sample container should be completely filled, tightly sealed, and delivered to the lab within 24 hours, ideally on ice in a cooler. The appropriate frequency of water sampling varies from once every 1 to 2 years to as often as quarterly, depending on the situation. In general, the deeper the well, the more slowly mineral concentrations are likely to change.

Where well water is impounded in a reservoir before use, the reservoir water should be tested because some constituents may change during impoundment. Irrigation districts typically have rigorous water testing programs, so the need for individual growers to test district water is limited to situations in which the grower receives water that has been mixed with return flows from upstream users.

RECORDKEEPING

Careful recordkeeping can enhance the accuracy and utility of a soil testing program. Errors can be made in the collection and analysis of samples, so checking results against the historical record can highlight questionable test values. Some soil characteristics can change rapidly (e.g., NO_3-N), so samples even a month apart may be quite different. However, in the absence of a high rate of fertilizer or amendment application, most nutrient availability characteristics change slowly; P, K, and micronutrient analysis fall into this category. A radical change in test results within 1 to 2 years may call into question the validity of those results and may justify retesting.

Tracking the change in soil values over time in conjunction with fertilizer and amendment application history and crop yields can help identify connections between soil fertility parameters, management practices, and crop productivity. This is particularly true for consultants or large growers who can develop a database consisting of many fields across years of production.

SOIL TEST METHODS

Soil Physiochemical Characteristics

pH and Salinity

Soil pH in the western United States is usually measured in a saturated paste made by mixing deionized water with dry soil until all the soil pore space has been filled. The pH is measured directly in the paste, while the electrical conductivity (EC, a measure of salinity) is measured in the clear supernatant collected by vacuum filtration of a saturated paste. Alternatively, a lab may measure pH and EC in an extract made by mixing dry soil 1:1 or 1:2 by weight with deionized water; this approach is faster and requires less labor. The pH results from 1:1 and 1:2 methods are generally very similar, with pH values from the saturated paste method being marginally lower. Unfortunately, it is difficult to directly compare EC results across methods, since these methods use differing ratios of soil and water, creating a dilution issue. Also, calcium compounds in soil are of low solubility; the higher the ratio of water to soil, the more calcium salts will be dissolved, influencing the EC measurement.

The concentration of individual salt constituents in a saturated paste extract may be measured to help diagnose problems with soil structure, water infiltration, or plant toxicity. Cations measured include Ca, Mg, Na, and occasionally K. The amount of these cations in a saturated paste extract represents only a small fraction of their overall soil availability. Anions commonly measured in saturated paste extracts include NO_3, SO_4, bicarbonate (HCO_3), Cl, and B.

Texture

The most common lab method for determining soil texture is the hydrometer method, in which the specific gravity of a soil water mixture is measured over time. Soil particles sink to the bottom of the container over time, lowering the specific gravity of the mixture. Sand particles drop out quickly, while clay particles remain in the water column for many hours; measuring specific gravity after the sand particles have dropped, and again after the silt particles have dropped, allows estimation of the percentage of soil weight represented by each particle size category. The proportions of sand, silt, and clay can be used to assess soil physical properties and water-holding capacity.

Organic Matter

Historically, a standard method to estimate soil organic matter was to measure the amount of organic carbon in a soil. However, this approach, called the Walkley-Black method, has fallen out of favor because it is complex and generates hazardous chemical waste. Most commercial labs have now adopted a method called loss on ignition (LOI), in which soil is heated in a muffle furnace to burn off organic matter. Measuring the change in soil weight after heating provides an estimate of the organic matter content. This analytical method tends to give estimates of soil organic matter that are 10 to 20 percent higher than the Walkley-Black method.

Cation Exchange Capacity

Cation exchange capacity (CEC) is properly measured using a complicated multistep process in which ammonium (NH_4^+), sodium (Na^+), or barium (Ba^{2+}) is used to displace the resident cations; the amount of NH_4^+, Na^+, or Ba^{2+} is then measured after displacement by a different cation. This test is typically not included in a standard soil test, although labs may perform this procedure for an extra fee. As a surrogate measurement, labs often report an estimated CEC, which is calculated from the results of the exchangeable cations test. The estimated CEC may be substantially different from the actual CEC. In acid soils, estimated CEC is usually lower than the actual CEC. In alkaline soils containing free lime (calcium carbonate), the exchangeable cations test solubilizes some Ca not on exchange sites; therefore, the estimated CEC can be significantly higher than the actual CEC. For soils with pH from 6.0 to 7.2, the estimated CEC should be close to the actual CEC.

Soil Fertility Tests

Lab methods used to estimate soil nutrient availability can be classified as either quantitative or index tests. A quantitative test measures an unambiguous characteristic that has an absolute value. Measurement of soil NO_3-N concentration is a quantitative test; the test procedures used are designed to remove and measure all NO_3-N ions in the soil. By contrast, the procedures used to measure P, K, and micronutrient availability are index tests. These nutrients have complex chemical interactions in soils, and no lab test can precisely measure the fraction of the total soil nutrient content that will be available to plants over a growing season.

This distinction between quantitative tests and index tests is important. One can use the results of a quantitative test to calculate how much of that nutrient is currently plant available. However, correct interpretation of an index test requires field calibration (evaluation of crop response to application of that nutrient across a range of soil test results) to establish interpretive guidelines. One cannot use the results of an index test to accurately calculate the amount of plant-available nutrient.

Nitrogen

Soils typically contain thousands of pounds of N per acre, but most of it is bound up in organic matter and is not directly available for crop uptake. Only NO_3-N and ammonium-nitrogen (NH_4-N) are immediately plant available. The concentration of NH_4-N is usually low in soil, so NO_3-N is the only N form commonly included in an agricultural soil test. Soil NO_3-N can be easily extracted from soil because as an anion, it is not held on soil particles. Labs use different extractants and analytical techniques to measure soil NO_3-N, but all tests should provide similar results.

Phosphorus

Similar to the situation with N, agricultural soils typically contain large amounts of phosphorus, but only a small fraction of it is plant available in a given growing season. Lab methods for assessing soil P availability attempt to rank a soil's ability to supply P over time. Two analytical procedures are commonly used by California soil testing labs, the Bray and Olsen tests. In the Bray test, soil is extracted with a weak acid solution; the Olsen test (also called the bicarbonate test) uses an alkaline solution (pH 8.5). The pH of the extracting solution affects the chemical forms of soil P that are dissolved, in some ways mimicking the pH effects of the soil solution in the field. Traditionally, the Olsen test was thought to be appropriate for soils above pH 6.0, while the Bray test was more appropriate for soils with pH below 6.0. However, recent research suggests that the Olsen test can provide useful information for soils with pH as low as 5.2 (Bair and Davenport 2012). This means that the Olsen procedure is appropriate for the vast majority of California soils. Correlation between the Bray and Olsen test values is poor due to the differences in extractant pH.

Potassium and Other Cations

Soil availability of K, Ca, and Mg is estimated by extracting soil with a concentrated solution of ammonium acetate. The NH_4^+ ions displace K, Ca, and Mg, and Na from soil cation exchange sites; this method is therefore referred to as the exchangeable cations test. However, not all the K and Ca measured by this procedure represent ions removed from cation exchange sites. A fraction of the extracted K (up to 25 percent in some soils) represents "fixed" K removed from between structural layers of certain types of soil particles (see chapter 6). In soil above pH 7.0, ammonium acetate extraction can dissolve some precipitated Ca compounds, inflating the exchangeable Ca value.

Micronutrients

Soil availability of Zn, Fe, Mn, and Cu is estimated by extraction with a solution of DTPA (diethylenetriaminepentaacetic acid, a chelating agent). As is the case with soil P, these micronutrients exist in various chemical compounds in soil with varying plant availability; the DPTA solution extracts the most active fraction. Soil Mo concentration is generally too low to be accurately measured by any practical lab technique. Therefore, Mo deficiency is typically identified through analysis of plant tissue. Mo deficiency is rare in California vegetable production.

In California soil, B concentration is more frequently excessive than it is deficient. Evaluation of B toxicity is done using saturated paste extraction. Where soil B deficiency is a possibility, either a hot water extraction or a DTPA-Sorbitol procedure is used.

Irrigation Water Test Procedures

Irrigation water testing is used to evaluate salinity, toxic ions, effects on soil structure, and water infiltration or to predict interactions with fertigated chemicals. The common constituents measured in irrigation water are the same as those measured in a soil saturated paste extract. Important nutrient constituents are NO_3-N and sulfate-sulfur (SO_4-S). The concentration of NO_3-N in groundwater is increasing across California, and many irrigation wells now contain enough NO_3-N to supply a significant portion of crop N requirements. The SO_4-S concentration of irrigation water varies greatly, but many wells and some surface waters also have ample SO_4-S to satisfy crop needs, irrespective of soil S availability. Irrigation water K concentration is usually too low to be agronomically significant. Water may contain large quantities of Ca and Mg, but since soil availability of these nutrients is almost always sufficient for crop productivity, the nutrient availability aspect of water Ca or Mg concentration is usually of minimal importance. However, the relative amount of Ca and Mg in irrigation water can have a significant effect on soil physical properties like surface crusting and water infiltration rate (see chapter 7).

Interpretation of Soil and Water Tests

Units and Conversions

Lab results may be reported in units that complicate interpretation. Soil and water constituents are usually measured on the basis of concentration, which can be expressed as parts per million (ppm), milligrams per kilogram (mg/kg), or milligrams per liter (mg/L); these are equivalent values (1 ppm = 1 mg/kg = 1 mg/L). However, some labs report values on soil test reports as pounds of available nutrient per acre-foot of soil or pounds per acre in the top 6 inches of soil. This conversion from concentration to pounds per acre can be misleading for several reasons. First, the conversion is based on the assumption that all soils have the same bulk density, which is not correct. The typical conversion factors used by labs (4,000,000 pounds of dry soil per acre-foot, or 2,000,000 pounds per acre in the top 6 inches) are actually higher than the weight of soil in most California fields. A more accurate conversion would assume a dry soil weight of 3,500,000 to 3,800,000 pounds per acre-foot for tilled mineral soils. Soils with very high organic matter content (greater than 10 percent, found in the Sacramento–San Joaquin Delta or Tulelake regions) are much lighter, typically weighing 2,500,000 to 3,000,000 pounds per acre-foot. Second, the lab tests for nutrient availability for all nutrients except NO_3-N are index tests, not quantitative tests, and as such, they cannot give correct results on the basis of pounds per acre.

When evaluating nutrient availability in soil or irrigation water it is necessary to understand the relationship between concentration (ppm or mg/L) and milliequivalents (meq, the number of positive or negative charges). Elements have different weights per unit of charge, and it is on the basis of charge that the relative effects of the various elements on crop nutrient uptake and soil physiochemical characteristics are best evaluated. Table 3.1 gives the numerical factors for converting between concentration and milliequivalents for both cations and anions in soil and water.

It is common for labs to report exchangeable cations as a concentration (ppm) and also as a percentage of base saturation or cation saturation. Percent saturation is the relative amount of the total cation exchange (in meq/100 g) represented by the cations. The difference between base saturation and cation saturation is that base saturation considers only the base cations (Ca^{2+}, Mg^{2+}, K^+, and Na^+), while cation saturation includes the concentration of H^+ as well. H^+ concentration is never measured but rather is estimated based on the soil pH. In soil above pH 6.5, the amount of H^+ is small, and base saturation and cation saturation are essentially the same. As soil pH gets progressively more acidic, H^+ represents a larger fraction of the cation exchange.

TABLE 3.1.

Numerical factors for conversion between units of cation concentration and charge

To convert column 1 into column 2, divide column 1 by	Column 1	Column 2	To convert column 2 into column 1, multiply column 2 by
	ppm exchangeable cations to/from meq/100 g of soil		
390	ppm K	meq K/100 g	390
200	ppm Ca	meq Ca/100 g	200
120	ppm Mg	meq Mg/100 g	120
230	ppm Na	meq Na/100 g	230
	ppm to/from meq/liter in saturated paste extracts or irrigation water		
39	ppm K	meq K/liter	39
20	ppm Ca	meq Ca/liter	20
12	ppm Mg	meq Mg/liter	12
23	ppm Na	meq Na/liter	23
14	ppm NO_3-N	meq NO_3-N/liter	14
16	ppm SO_4-S	meq SO_4-S/liter	16
35	ppm Cl	meq Cl/liter	35

Percent base saturation can be calculated using the conversion factors in table 3.1. For example, if a soil has exchangeable Ca, Mg, K, and Na concentrations of 2,200, 480, 390, and 460 ppm, respectively, the milliequivalents of each cation would be

2,200 ppm Ca ÷ 200 = 11 meq Ca/100 g

480 ppm Mg ÷ 120 = 4 meq Mg/100 g

390 ppm K ÷ 390 = 1 meq K/100 g

460 ppm Na ÷ 230 = 2 meq Na/100 g.

Therefore, the percent base saturation of Ca would be

(11 meq Ca ÷ 18 total meq of base cations) x 100 = 61%.

EC is measured in units related to the conduction of electricity (higher salt concentration equals greater conductivity). Even here the units reported can be confusing. Most labs now report EC in decisiemens per meter (dS/m), but some labs still use millimhos per centimeter (mmhos/cm) or even millisiemens per centimeter (mS/cm). All of these units have the same numerical value: 1 dS/m = 1 mmhos/cm = 1 mS/cm.

Accuracy of Lab Analysis

Lab analysis contains numerous potential sources of inaccuracy. Some analyses are more complex than others and therefore are more prone to variability in measurement. When evaluating soil or water reports, recognize that some degree of uncertainty exists in the results. In general, a well-managed lab will consistently report values within about 10 percent of the correct value. This means that the correct value for a reported Olsen P of 20 ppm is likely to be from about 18 to 22 ppm. Soil pH is an exception, with labs generally reporting values ± 0.2 units of the correct value. Most agricultural testing labs in California participate in one or more certification programs designed to document that they meet industry performance standards, and many labs are willing to share that documentation with prospective clients.

Interpreting Saturated Paste and Irrigation Water Results

Soil pH

Figure 3.1 shows the effect of pH on soil nutrient availability. No pH maximizes the availability of all nutrients. The availability of the base cations (Ca, Mg, and K) is maximized in alkaline soils, while most micronutrients are maximized in mildly acidic soils. However, practical experience has shown that the pH sweet spot for overall soil nutrient availability is approximately 6.0 to 7.5. Within this range most vegetable crops grow well, and pH-induced nutrient limitations are generally a minor concern. Many crops tolerate an even wider pH range: for example, California's highest tomato and melon yields are achieved in the Westside region of the San Joaquin Valley where many fields have pH over 7.5.

Soil pH influences more than just nutrient availability. The severity of some soilborne diseases is affected by soil pH, and pH ma-nipulation can be a viable control strategy. Examples include maintaining pH above 7.2 to suppress clubroot in *Brassica* spp. crops and maintaining pH below 6.0 to reduce common scab of potato.

Saturation Percentage

In addition to pH, labs often report the saturation percentage (SP), which is the ratio of water to dry soil in a saturated paste expressed as a percentage. Several important soil properties can be inferred from SP. First, the gravimetric soil water content at field capacity can be estimated by dividing SP by 2 because at field capacity the pore spaces in soil are about evenly split between water and air. Second, one can estimate soil texture from SP (table 3.2): as the amount of silt and clay in a soil increases, the SP increases. This relationship is confounded in soils with higher levels of organic matter because soil organic matter holds much more water on a dry weight basis than do mineral soil particles.

Salinity

Research at the USDA Salinity Lab in Riverside, California, documented the relative tolerance of many crops to salinity in soil and irrigation water. Based on that work, table 3.3 ranks the salinity tolerance of common vegetable crops. Since soil salinity can change substantially over time, one must consider the timing of sample collection when interpreting the EC of a saturated paste extract (EC_e). Soil samples collected in the fall before winter rains often have much higher salinity than those taken in the spring. Irrigation water salinity is generally quite stable over time. Salinity can also be reported as total dissolved solids (TDS). Conversion between EC and TDS can be done, but it is an approximation because the relationship is affected by the salinity level and the composition of salts in the water. The generic conversions are

$$EC \text{ (dS/m)} \times 640 = TDS \text{ (ppm)},$$
$$\text{when EC is} < 5 \text{ dS/m}$$

or

$$EC \text{ (dS/m)} \times 800 = TDS \text{ (ppm)},$$
$$\text{when EC is} > 5 \text{ dS/m}.$$

In addition to the total salt load in soil, the salt composition can have important

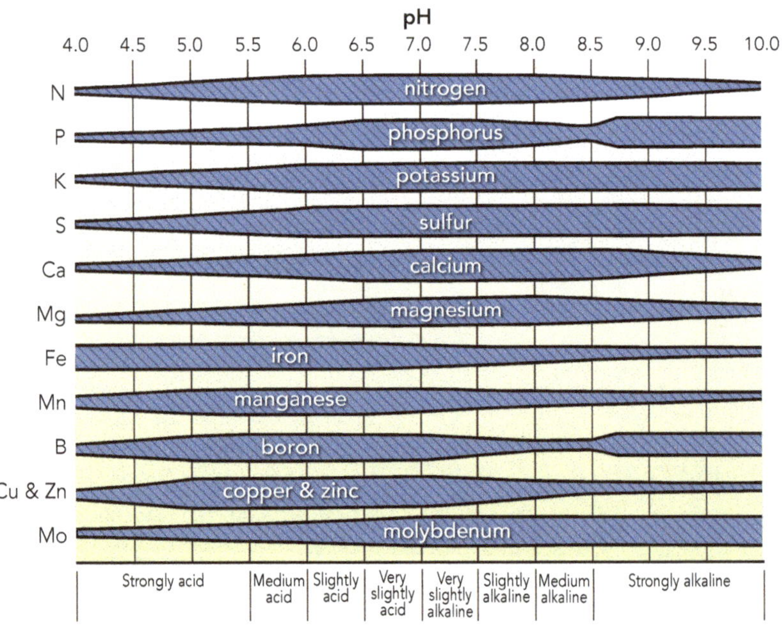

FIGURE 3.1.

Effect of soil pH on relative nutrient availability.

TABLE 3.2.

Relationship between saturation percentage and soil texture

Saturation percentage	Soil texture
less than 20	sand or loamy sand
20–35	sandy loam
35–50	loam or silt loam
50–65	clay loam
more than 65	clay, or organic soil

Source: Adapted from Robbins and Wiegand 1990.

TABLE 3.3.

Electrical conductivity (EC) thresholds (in dS/m) above which crop yield potential is reduced; thresholds are for saturated paste soil extracts (EC$_e$) or irrigation water (EC$_w$)

EC$_e$ < 1.5 or EC$_w$ < 1.0	EC$_e$ 1.5 to 2.5 or EC$_w$ 1.0 to 1.7	EC$_e$ > 2.5 or EC$_w$ > 1.7
bean	celery	beet
carrot	lettuce	broccoli
lettuce	potato	squash
onion	spinach	
pepper	tomato	

Sources: Adapted from Grattan 2002; Maas 1984.

TABLE 3.4.

Irrigation water thresholds (in ppm and meq/L) for sodium and chloride toxicities in vegetable crops

		Degree of restriction on use		
		None	*Slight to moderate*	*Severe*
sodium	ppm	< 115	115–460	> 460
	meq/L	5	5–20	> 20
chloride	ppm	< 175	175–700	> 700
	meq/L	5	5–20	> 20

Source: Adapted from Ayers and Westcot 1985.

effects. Sodium can be toxic to plants at high concentration and can have adverse effects on soil structure and water infiltration. Soil Na can be measured in two ways: as the sodium adsorption ratio (SAR) or as the exchangeable sodium percentage (ESP). SAR is calculated using the relative amounts of Na, Ca, and Mg (in milliequivalents) in a saturated paste extract and the formula

$$\text{SAR} = \text{Na} \div [\text{square root } ((\text{Ca} + \text{Mg}) \div 2)].$$

ESP is calculated from the results of the exchangeable cations test using the formula

$$\text{ESP} = (\text{meq Na} \div \text{meq of estimated CEC}) \times 100.$$

An ESP greater than 15 or an SAR greater than 13 indicates a serious sodium hazard. An ESP or SAR below 5 indicates no sodium-related restrictions (Horneck et al. 2007).

Adverse effects of irrigation water Na on soil structure and water infiltration depend on the EC of the water: the greater the EC, the higher SAR can be without inducing adverse soil effects. In general, a SAR less than 6 is problematic only with water of very low EC (less than 0.3 dS/m), while a SAR greater than 12 represents an infiltration problem in most circumstances. More detail on sodium effects on soil structure is given in Grattan (2002).

The potential for crop toxicity from Na, Cl, and B is usually rated in irrigation water instead of soil. Table 3.4 lists general guidelines for assessing the Na and Cl hazard of irrigation water. The hazard is increased with sprinkler irrigation because these salts can concentrate on leaf tips as they dry. Boron tolerance varies widely among vegetable crops; irrigation water with 1 ppm B is tolerated by most vegetables, while water above 4 ppm B damages most crops (Maas 1986). Additional detail on B sensitivity is given in chapter 8.

Nutrient Content

While testing of saturated paste extracts and irrigation water is usually associated with salinity or toxicity issues, some nutrient-related information can also be gleaned. Sulfate-sulfur (SO_4-S) in excess of 1 milliequivalent per liter (16 ppm S) in a saturated paste extract suggests substantial S availability. However, SO_4-S concentration in a saturated paste extract can be low if the sample was taken after winter rains, so a low concentration is not definitive as to the sulfur fertility in that field. Irrigation water with 1 milliequivalent per liter SO_4-S contains 3.6 pounds of sulfur per acre-inch, so water at or above this level contributes significantly to crop S requirements.

Nitrate-nitrogen concentration might not be reported from a saturated paste extract. If it is reported, convert the extract concentration to concentration on a dry soil basis using this formula:

$$\text{ppm saturated paste } NO_3\text{-N} \times (\text{saturation percentage} \div 100) = \text{ppm } NO_3\text{-N in dry soil}$$

For example, a saturated paste extract of 20 ppm NO_3-N with a 40 percent saturation percentage would have 8 ppm NO_3-N on a dry soil basis.

Irrigation water analysis typically includes NO_3-N concentration. The amount of NO_3-N in irrigation water can be calculated as follows:

$$\text{ppm } NO_3\text{-N} \times 0.23 = \text{lb } NO_3\text{-N per acre-inch of water}$$

$$\text{ppm } NO_3\text{-N} \times 2.72 = \text{lb } NO_3\text{-N per acre-foot of water}$$

Some labs report nitrate concentration on the basis of the NO_3. ion; the conversion to NO_3-N is:

$$\text{ppm } NO_3 \div 4.43 = \text{ppm } NO_3\text{-N.}$$

Chapter 10 describes how to estimate a fertilizer credit for irrigation water NO_3-N.

Interpreting Soil Fertility Results

Index tests of soil nutrient availability require field calibration to establish interpretive guidelines. Field calibration involves measuring crop response to fertilization with a particular nutrient across a wide range of soil test values for that nutrient, then using these data to construct an interpretation guide (fig. 3.2). Extensive soil P and K calibration data exists for some major field crops, but for many of California's important horticultural crops little data is available from public sources. Similarly, calibration data on soil micronutrient availability is very limited. Consequently, individual testing labs may use somewhat different interpretive standards, as they may be drawing data from other regions or from proprietary databases.

Nitrogen

Soil NO_3-N concentration is a quantitative measure of current soil N availability. In California agriculture, particularly in annual crop rotations, understanding soil NO_3-N testing and interpretation is critical to efficient N management. Factoring soil test NO_3-N into an N fertilization plan can be complex and requires consideration of individual field circumstances. However, generalizations can be made. Soil NO_3-N below 5 ppm indicates low N availability. Values from 10 to 20 ppm NO_3-N are common in fertilized root zones and indicate that short-term crop N uptake can be satisfied. Soil NO_3-N above 20 ppm suggests soil N availability is high enough to supply actively growing crops for an extended period, and additional N application may be postponed or reduced to reflect the current soil N supply. Chapter 9 contains an in-depth discussion of integrating soil NO_3-N test information into N fertilization management.

Soil organic matter and total soil N content are closely correlated; in general, soil with higher organic matter content is likely to

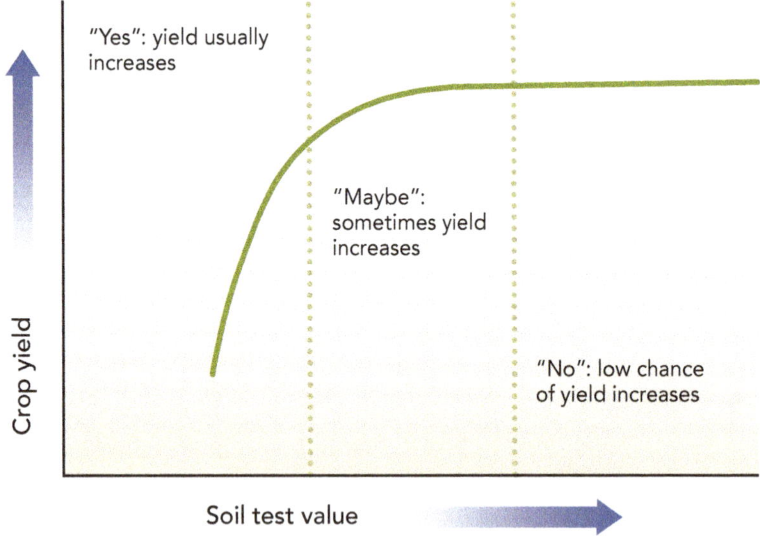

FIGURE 3.2.

Relationship between soil test value and the likelihood of achieving a crop yield response to fertilization with a given nutrient.
Source: Adapted from Horneck et al. 2011.

mineralize and release more plant-available N over time. Consequently, some labs report in soil test results a value for "estimated N release" or a similar category. This value is not based on a separate test procedure but rather is calculated based on soil organic matter. However, the correlation between soil organic matter content and the amount of N that soil mineralizes in a growing season is weak. Actual in-season soil N mineralization is affected by many factors, including crop rotation, amendment history, tillage and irrigation management, and other factors. This topic is covered more comprehensively in chapter 4.

Phosphorus

Correct interpretation of soil P test results requires consideration of the crop to be grown and the soil temperature range the crop will encounter, because crops vary greatly in their P uptake requirements and in their efficiency in removing P from soil solution. Also, soil P is more available in warm soil (above 70°F) than in cool soil (below 60°F). Lettuce and celery require relatively high soil P availability for maximum growth, while the growth of tomatoes and cucurbits can be maximized at a lower soil P level. Table 3.5 contains soil test P interpretation guidelines. Only Olsen test values are given, since the Olsen test is the most appropriate soil test for the vast majority of California soils (those above about pH 5.5). These guidelines are based on a sample depth of 12 inches. Because soil P availability generally declines with depth, soil samples taken only 6 inches deep may substantially overestimate soil P availability, and these interpretation values should be adjusted accordingly.

Some labs give both Olsen and Bray P results, which can be quite different numerically. Bray P extraction is useful only in acidic soils (below about pH 5.5), and there is insufficient California calibration data to provide crop-specific interpretation of Bray P results. Bray P categories for western Oregon have been suggested to be less than 20 ppm, low; 20 to 40, medium; and greater than 100, excessive (Horneck et al. 2011). This rating scale is not crop specific, so the categories may need adjustment for crops with a high soil test P response threshold. Additional detail is given in chapter 5.

Potassium

Vegetable crops vary widely in K uptake and therefore have different soil K requirements. In addition to the crop being grown, other factors affecting soil test K interpretation are soil physical characteristics, the relative abundance of other soil cations, and irrigation practices. Rooting density greatly affects a plant's ability to extract K from the soil; any soil characteristic (subsurface compaction, poor structure or aeration, etc.) that limits rooting density or depth restricts K uptake. Other soil cations can compete for plant cation uptake; therefore, soil K availability should be evaluated both on a concentration (ppm) basis and as a percentage of exchangeable soil cations. Cation competition is generally not a significant issue in soil in which K makes up more than 3 percent of exchangeable cations (on a milliequivalent basis), whereas soil in which K makes up less than 2 percent of exchangeable cations may have restricted K availability, even with

TABLE 3.5.

Probability of a crop response to P fertilization based on the Olsen (bicarbonate-extractable) soil P level

Crop	Crop response likely*	Response possible**	Response unlikely*
	Olsen P (ppm)		
lettuce and celery	< 40	40–60	> 60
other cool-season vegetables	< 25	25–35	> 35
warm-season vegetables (tomato, pepper, potato, cucurbits)	< 15	15–25	> 25

*Regardless of soil temperature.

**Response more likely in soils cooler than 60°F.

relatively high exchangeable K concentration. Any irrigation practice that concentrates rooting in a small area (i.e., drip irrigation) may restrict soil K uptake more than a practice that wets a larger soil volume. Table 3.6 suggests generalized guidelines based on exchangeable soil K concentration. These guidelines should be adjusted based on the other characteristics discussed. Additional detail is given in chapter 6.

Calcium and Magnesium

Nearly all California soils contain sufficient Ca and Mg to meet the nutritional requirements of all vegetable crops. Crop uptake requirements are typically less than 100 pounds Ca and 50 pounds Mg per acre, while most fields have hundreds or often thousands of pounds of exchangeable Ca and Mg in the top foot of soil. Exceptions may be very sandy soils (very low CEC) and strongly acidic soils (H^+ ions and Al^{3+} displace Ca and Mg on the cation exchange sites); such soils are rare in California. Physiological disorders associated with Ca supply such as blossom end rot of tomato and black heart of celery do occur, but the cause is seldom a lack of available soil Ca. This apparent contradiction is explained in chapter 7.

In soils with neutral to alkaline pH, Ca typically accounts for 40 to 85 percent of exchangeable cations on a milliequivalent basis, while Mg accounts for 10 to 50 percent. Significant regional differences exist in the relative amounts of Ca and Mg in soils. The Sacramento Valley has relatively high soil Mg, a legacy of the serpentine (high-Mg) parent material from which the soil formed. Conversely, most soils in the San Joaquin Valley, Southern California, and the Central Coast valleys are calcium dominated. Within regions, individual fields differ substantially in cation balance; using calcium-containing amendments (lime, gypsum) and irrigation water with cation content can substantially affect the balance between soil Ca and Mg over time.

While soil Ca and Mg seldom limit crop growth, the balance between them should be

TABLE 3.6.

Probability of a crop response to K fertilization based on the ammonium acetate exchangeable soil K level

	Crop response likely*	Response possible**	Response unlikely*
Crop	Exchangeable K (ppm)		
celery	< 150	150–200	> 200
other cool-season vegetables	< 100	100–150	> 150
potato, tomato, pepper	< 150	150–200	> 200
cucurbits	< 80	80–120	> 120

TABLE 3.7.

Interpretation of soil test results for DTPA-extractable micronutrients

		Crop response likely*	Response possible**	Response unlikely*
Element	**Symbol**	DTPA-extractable micronutrients (ppm)		
zinc	Zn	< 0.5	0.5–1.5	> 1.5
iron	Fe	< 5.0	5.0–10.0	> 10.0
manganese	Mn	< 1.5	1.5–5.0	> 5
copper	Cu	< 0.5	0.5–2.0	> 2.0
		Hot water extraction (ppm)		
boron	B	< 0.2	0.2–1.0	> 1.0

Sources: Brown and deBoer 1983; Havlin et al. 2014; Horneck et al. 2011; Ludwick et al. 2010.

evaluated. Soil structure and water infiltration are usually better in soils with a higher Ca to Mg ratio. Soil Ca to Mg ratio is best evaluated using a saturated paste extract, because in alkaline soils the exchangeable cations test may overestimate Ca availability. As a general rule, a saturated paste Ca to Mg ratio of 2:1 (milli-equivalent basis) or higher is desirable for soil structure and water infiltration, with ratios above 2:1 conferring little additional benefit. Soils with a Ca to Mg ratio approaching 1:1 are likely to have limited water infiltration. As a practical matter, substantially adjusting the soil Ca to Mg ratio of the entire soil profile can be economically prohibitive, requiring many tons of calcium-containing amendments per acre. For this reason, calcium amendment strategies to improve soil physical properties often concentrate on only the top few inches of soil. Other practices such as cover cropping and applying organic amendments can achieve some of the same soil structure benefits as calcium application.

Micronutrients

Yield-limiting soil micronutrient deficiency is uncommon in California vegetable production. Consequently, relatively little research on soil micronutrient issues has been done in recent decades, and there is limited California data upon which to base soil test interpretive standards. While acknowledging this limitation, interpretive standards are suggested in table 3.7. Crops differ widely in their ability to extract micronutrients from the soil, and the likelihood of deficiency is greater in some vegetables than in others (see chapter 8). However, even for sensitive crops, the soil test values in the "response unlikely" range in table 3.7 should be adequate for good production. Where micronutrient deficiency is a possibility, in-season plant tissue analysis may be warranted in addition to soil testing.

REFERENCES

Ayers, R. S., and D. W. Westcot. 1985. Water quality in agriculture. FAO irrigation and drainage paper 29. Rome: U.N. Food and Agriculture Organization.

Bair, K. E., and J. R. Davenport. 2012. Influence of recent acidification on available phosphorus indices and sorption in Washington State soils. Soil Science Society of America Journal 76:515–521.

Brown, A. L., and G. J. deBoer. 1983. Soil tests for zinc, iron, manganese and copper. In H. M. Reisenauer, ed., Soil and plant tissue testing in California. Oakland: University of California Agriculture and Natural Resources Publication 1879.

Grattan, S. R. 2002. Irrigation water salinity and crop production. Oakland: University of California Agriculture and Natural Resources Publication 8066.

Havlin, J. L., S. L. Tisdale, W. L. Nelson, and J. D. Beaton. 2014. Soil fertility and fertilizers. Upper Saddle River, NJ: Pearson.

Horneck, D. A., J. W. Ellsworth, B. G. Hopkins, D. M. Sullivan, and R. G. Stevens. 2007. Managing salt-affected soils for crop production. Corvallis: Pacific Northwest Cooperative Extension Publication PNW 601-E.

Horneck, D. A., D. M. Sullivan, J. S. Owen, and J. M. Hart. 2011. Soil test interpretation guide. Corvallis: Pacific Northwest Cooperative Extension Publication EC 1478.

Lazcano, C., J. Wade, W. R. Horwath, and M. Burger. 2015. Soil sampling protocol reliably estimates preplant NO_3- in SDI tomatoes. California Agriculture 69:222–229.

Ludwick, A. E., L. C. Bonczkowski, M. H. Buttress, C. J. Hurst, S. E. Petrie, I. L. Phillips, J. J. Smith, and T. A. Tindall. 2010. Western fertilizer handbook. 9th ed. Long Grove, IL: Waveland Press.

Maas, E. V. 1984. Crop tolerance. California Agriculture 38(10): 20–21.

———. 1986. Salt tolerance of plants. Applied Agricultural Research 1:12–26.

Robbins, C. W., and C. L. Wiegand. 1990. Field and laboratory measurements. In K. K. Tanji, ed., Agricultural salinity, assessment, and management. ASCE Manuals and Reports 71. New York: American Society of Civil Engineers. 201–219.

Nitrogen Management

N itrogen management is a key issue in vegetable crop production. Historically, vegetable growers relied on heavy N fertilization to ensure peak crop yields and quality. In recent years, N management practices have come under intense scrutiny by regulatory agencies concerned about agriculture's contributions to N pollution of groundwater, surface water, and the atmosphere. This regulatory pressure will require growers to manage N more efficiently.

Efficient N management requires an in-depth understanding of the interplay of physical, chemical, and biological processes that occurs in agricultural soils. Figure 4.1 illustrates the important features of the agricultural soil N cycle. In this cycle, N is added to soil mainly through fertilizer application, use of organic amendments, and as nitrate in irrigation water. Nitrogen leaves a field in harvested products, in gases lost to the atmosphere, and as nitrate in surface runoff, tile drainage, and deep leaching. Nitrogen applied to a field from any source that is not removed from the field in a harvested product is theoretically at risk of being lost to the environment over time. Nitrogen balance (N applied minus N removed) is likely to be the main metric that regulatory agencies will use to evaluate the N efficiency of grower practices.

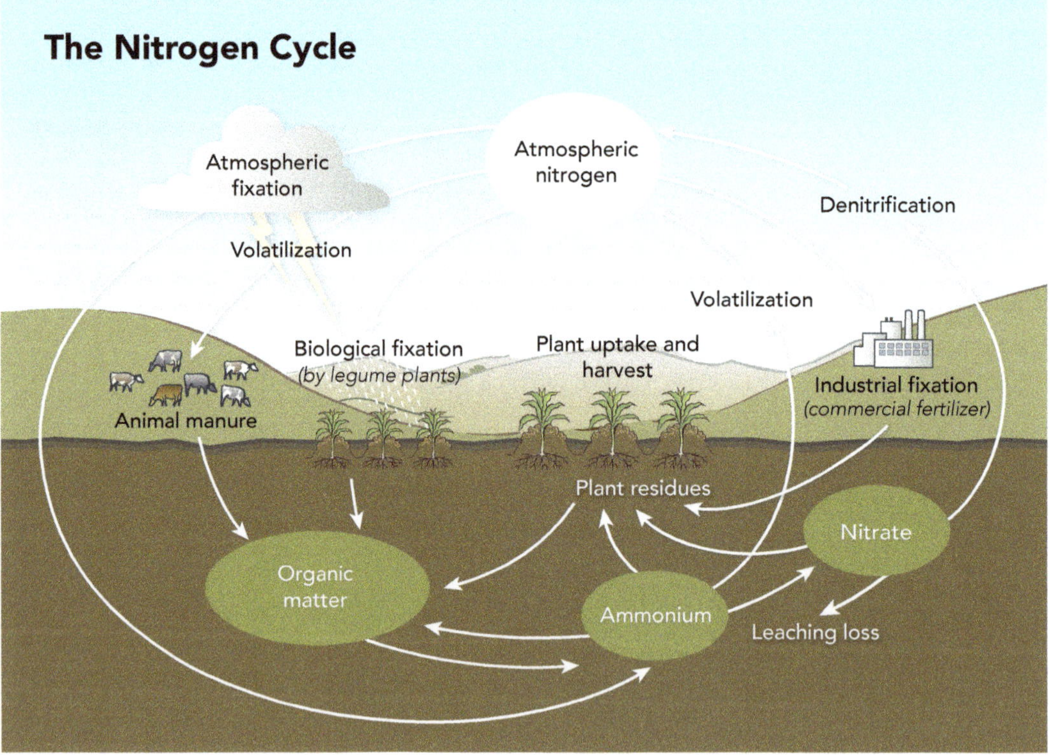

FIGURE 4.1.

The agricultural N cycle.
Source: The Fertilizer Institute, tfi.org.

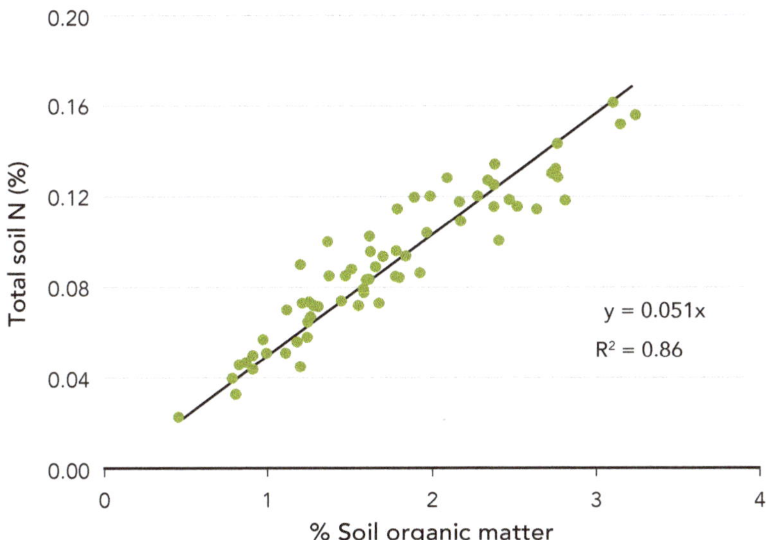

FIGURE 4.2.

Relationship between soil organic matter (by loss on ignition method) and total soil N for representative California soils.
Source: Hartz unpublished data.

SOIL NITROGEN FORMS AND TRANSFORMATIONS

Plants take up the overwhelming majority of their N in mineral form as NO_3-N or NH_4-N, yet mineral N makes up only a tiny fraction of the total N in soil. The vast majority of soil N is bound up in organic material and is not readily available for crop uptake. Nitrogen typically makes up approximately 5 percent of soil organic matter (fig. 4.2). This means that the top foot of a mineral soil usually contains 1,500 to 2,000 pounds of N per acre for each percent of soil organic matter; a soil with 2 percent organic matter contains roughly 3,000 to 4,000 pounds of N per acre-foot. This relationship is somewhat different for soils with high organic matter (peat) like those in the Sacramento–San Joaquin Delta and Tulelake regions, where organic matter can exceed 20 percent. A somewhat lower N concentration in soil organic matter, combined with much lower soil bulk density, means that the organic N content of peat soils typically averages from 1,000 to 1,500 pounds of N per acre-foot for each percent organic matter. A peat soil with 10 percent organic matter typically contains from 10,000 to 15,000 pounds of organic N per acre-foot.

Soil organic matter consists of all materials of biological origin in the soil. These include living microorganisms, actively decomposing organic material (crop residue or organic amendments incorporated in recent years), and a stable fraction called humus (fig. 4.3). Humus is relatively resistant to breakdown, so it has relatively little effect on soil N supply. However, it exerts substantial influence on other important soil properties (increasing CEC, improving soil structure and soil water retention, etc.). The living fraction of soil organic matter, also referred to as the soil microbial biomass, varies considerably over time based on the availability of easily decomposable organic material (crop residue, manure, compost, etc.). Microbial biomass, made up largely of bacteria and fungi, increases when organic matter is incorporated and declines once the easily degradable material has been broken down.

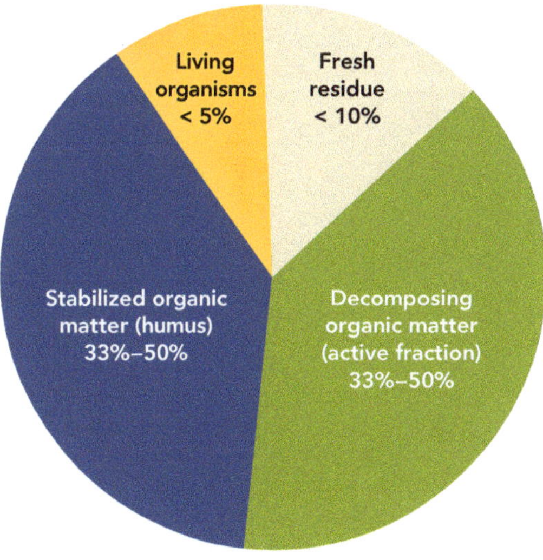

FIGURE 4.3.

Composition of soil organic matter.

Mineralization and Immobilization

Mineral N (NH_4^+, NO_3^-) is generated through the degradation of organic matter by soil microorganisms in a process called mineralization. The rate of mineralization is governed by soil temperature, soil moisture, and the decomposability and relative abundance of C and N (the C:N ratio) in the organic material being broken down. The N mineralization rate increases with temperature and is maximized when soil moisture is near field capacity. As a general rule, the mineralization rate of organic material with a C:N ratio less than 15:1 will be relatively rapid, while material with a C:N greater than 20:1 may temporarily decrease the amount of soil mineral N through the process of immobilization (Kumar and Goh 2000). Immobilization occurs when soil microbes multiply in response to the addition of organic material; if that material does not contain sufficient N to support microbial growth, the microbes will incorporate mineral N already present in the soil into their biomass. Immobilization may be a short-lived phenomenon or may persist for several months, depending on the amount of organic material incorporated and its C:N ratio.

Understanding the dynamics of mineralization and immobilization is critical to efficient N management. In vegetable rotations, large amounts of residue can be returned to the soil, and that residue usually contains from 2.5 to 5.0 percent N on a dry weight basis. Since crop residues typically contain about 40 percent carbon by dry weight, vegetable residues usually have a C:N ratio below 15:1. Therefore, N mineralization begins immediately following soil incorporation of vegetable residue, assuming that soil moisture is adequate to allow microbial activity. Laboratory incubation studies using vegetable residues have shown that about 30 to 60 percent of the residue N content is likely to mineralize within 8 weeks following incorporation (fig. 4.4); the higher the N concentration (and the lower the C:N ratio), the faster N mineralizes. Most of that mineralization occurs in the first 2 to 4 weeks after incorporation, with the rate of additional mineralization declining rapidly thereafter. Depending on the crop involved, this can represent a large amount of N. See table 2.1 for differences among crops in the amount of N contained in crop residue.

Residues of field crops (corn, cotton, wheat, etc.) grown in rotation with vegetables typically have much higher C:N ratios (from 20:1 to 80:1 is common) and therefore may temporarily induce immobilization. It is a common practice to apply some fertilizer N when incorporating residues with high C:N levels to stimulate residue breakdown and minimize the duration of the immobilization period. A rough estimate of the immobilization potential of residue with high C:N can be calculated using this equation (adapted from Havlin et al. 2014):

N immobilization (lb N/acre) = (lb dry residue × 0.0175) − [(lb dry residue × 0.4) ÷ C:N ratio]

For example, 4,000 pounds per acre of dry wheat straw with a C:N ratio of 50 would have an immobilization potential of about 38 pounds of N per acre. Because soil always contains some mineral N, not all this estimated immobilization must be supplied with applied N. As a general rule, applying 5 pounds of N per 1,000 pounds of dry biomass of residue with a high C:N ratio should be adequate to overcome immobilization. Estimates of the typical amount and C:N ratio of field crop residues are given in Mitchell et al. (1999).

Nitrogen mineralization from soil organic matter that occurs during a cropping season can also be an important source of crop nutrition. A number of studies have suggested that from 1 to 2 percent of soil organic N in the top foot of soil is likely to mineralize in a growing season under summer conditions (Castro and Hartz 2016; Hartz et al. 2000; Krueskopf et al. 2002). Using the rule of

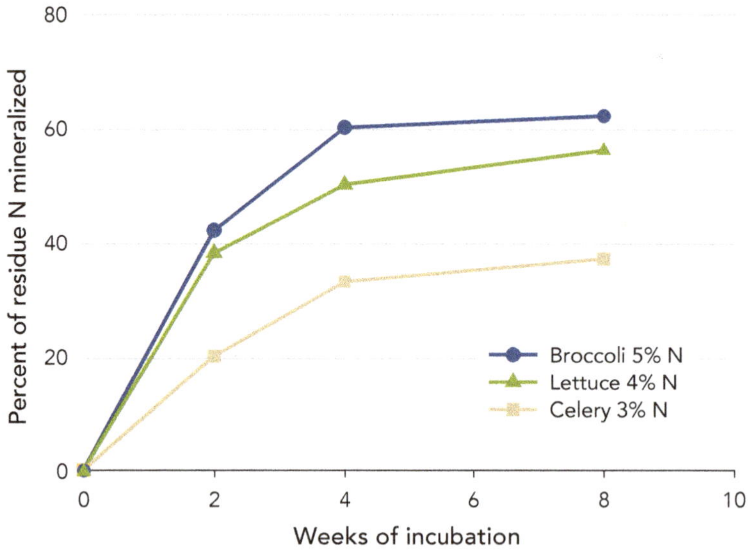

FIGURE 4.4.

Pattern of net N mineralization from vegetable crop residue incubated in an agricultural soil at 68°F.
Source: Hartz unpublished data.

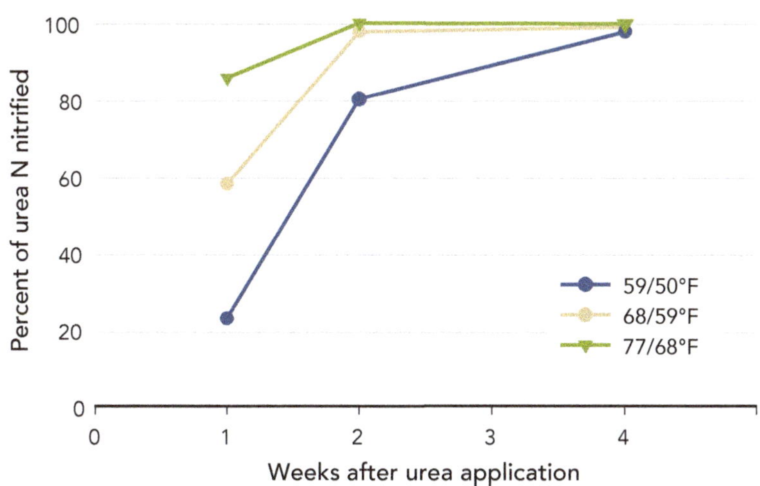

FIGURE 4.5.

Effect of soil temperature on the speed of nitrification of N applied to an agricultural soil as urea.
Source: Adapted from Agehara and Warncke 2005.

thumb that 1 acre-foot of soil contains at least 1,500 pounds of organic N for every percent of organic matter, a mineral soil with 2 percent organic matter may mineralize from 30 to 60 pounds of N per acre in a growing season. However, it is easy to see that soil management practices (tillage methods, use of organic amendments) and irrigation methods could have a substantial influence on N mineralization. For example, soils receiving repeated applications of manure compost may have a larger "active" fraction of soil organic matter and may therefore show more rapid mineralization. Fields that are sprinkler irrigated may show more mineralization than those irrigated by buried drip irrigation, because sprinklers keep a greater volume of soil moist. While it is appropriate to consider the contribution of in-season soil N mineralization when developing an N fertility plan, the uncertainties involved suggest that a conservative estimate should be used.

The calculation is somewhat different when estimating the N mineralization potential of peat soils like those common in the Delta or Tulelake regions. Not only do peat soils contain less N for every percent of soil organic matter, the overwhelming majority of that N is contained in the humus fraction of soil organic matter, which is more resistant to mineralization. Based on limited California data, a reasonable estimate is that from 0.5 to 1 percent of soil organic N is likely to mineralize in a cropping season. Therefore, a peat soil with 15 percent organic matter containing 1,200 pounds of soil N per acre-foot for every percent soil organic matter would be expected to mineralize from 90 to 180 pounds of N per acre in a growing season. Given the relatively large uncertainty associated with this estimate, in-season soil nitrate monitoring to guide N management (see chapter 9) could be a particularly useful practice in soils with high levels of organic matter.

Nitrification and Denitrification

As soil organic N is broken down in the process of mineralization, NH_4-N is the initial mineral N form liberated; NH_4-N is then relatively quickly converted to NO_3-N in a process called nitrification. Nitrification is accomplished by soil bacteria in a two-stage process. The initial conversion of NH_4^+ to nitrite (NO_2^-) is mediated by ammonia-oxidizing bacteria (e.g., *Nitrosomonas* spp.); the conversion of NO_2^- to NO_3^- is accomplished by other species of bacteria, such as *Nitrobacter*. The rate of nitrification increases with increasing temperature and soil pH and is maximized near field capacity soil moisture content. Therefore, conditions in California vegetable fields generally favor rapid nitrification, and the concentration of NH_4-N in soil is low (typically below 2 ppm). The exceptions to this rule occur after crop residue has been incorporated, when for a brief period the rate of mineralization may exceed the rate of nitrification, or shortly after an application of NH_4-N fertilizer.

When fertilizer containing NH_4-N or urea ($CO(NH_2)_2$, which breaks down quickly into NH_4-N) is incorporated into moist soil, the majority of the applied N is likely to be nitrified within 1 to 2 weeks. Figure 4.5 shows the rate at which N applied to an agricultural soil as urea was nitrified in a laboratory incubation; the temperature treatments simulated a range of day/night fluctuations in soil temperature. To put this information into context, mean monthly soil temperatures for representative California production areas are shown in figure 4.6. It is clear that under summer

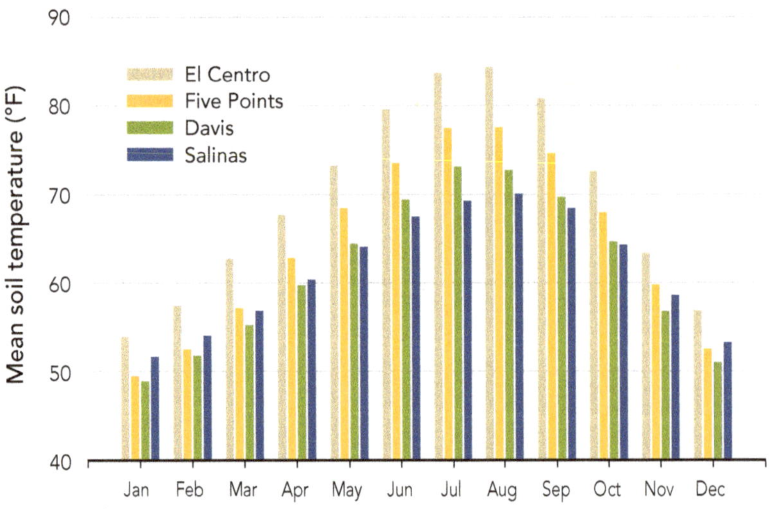

FIGURE 4.6.

Mean monthly soil temperature (6-inch depth) in major vegetable production areas.
Source: CIMIS network.

conditions the vast majority of applied NH_4-N or urea would be expected to nitrify within a week; even under winter conditions at least half would be expected to nitrify within 2 weeks. This suggests that the form of fertilizer N applied has only a temporary influence; under most field conditions the overwhelming majority of N taken up by a crop is likely to be in nitrate form, regardless of what fertilizer N source was applied.

In addition to being able to increase soil NO_3-N through nitrification, soil microbes can also reduce soil NO_3-N through the process of denitrification, in which NO_3-N is converted to various gaseous N forms, predominately N_2 (the nonreactive gas that makes up nearly 80 percent of the atmosphere). Denitrification is a multistep process mediated by soil bacteria that are only active under anaerobic (low-oxygen) conditions. In most agricultural situations, denitrification is a transient phenomenon, occurring briefly after a saturating irrigation or rain, with agronomically insignificant amounts of NO_3-N lost. However, in situations where all factors favor denitrification (high soil carbon availability due to multiple crops per year, heavy irrigation or rain, slowly draining soil, and high soil NO_3-N concentration), losses can be much higher, more than 100 pounds of N per acre per year (Ryden and Lund 1980). The

potential for denitrification tends to be greater in coastal production, where multiple crops per year are grown, sprinklers are used, and soil NO_3-N concentration tends to be high, than in drip-irrigated Central Valley production.

Nitrous oxide (N_2O) gas is an intermediate compound formed during nitrification and denitrification that can escape to the atmosphere. Release of N_2O is an environmental issue because it is a potent greenhouse gas, with approximately 300 times the activity of an equivalent amount of CO_2. While the emission of N_2O is limited to a few pounds per acre per production season in most field situations, this represents a significant portion of all greenhouse gas emissions from agriculture. In the future, N_2O may be regulated by the California Air Resources Board, which would provide an incentive to manage fields to limit denitrification.

Urea Hydrolysis

Urea, $CO(NH_2)_2$, is broken down to form NH_4^+ in an enzymatic process called hydrolysis. All agricultural soils contain the urease enzyme, and differences among soils in the rate of hydrolysis are relatively minor. Hydrolysis occurs rapidly when urea is applied below the surface of moist soil, as with banding, incorporation, or fertigation. The rate of hydrolysis increases with soil temperature (fig. 4.7); at soil temperatures typical of California conditions, hydrolysis is usually complete within several days after soil application. However, hydrolysis can be much slower when urea is broadcast on the soil surface without incorporation by tillage or irrigation.

Ammonia Volatilization

Loss of gaseous NH_3 (ammonia) to the atmosphere is called ammonia volatilization. In a liquid solution, an equilibrium exists between NH_3 and NH_4^+. This equilibrium favors NH_4^+ at lower pH and NH_3 at higher pH (fig. 4.8). In normal circumstances, both the soil pH and the soil solution NH_4-N concentration are low enough that ammonia volatilization is minimal. However, the application of urea or anhydrous ammonia dramatically increases the potential for volatilization. When anhy-

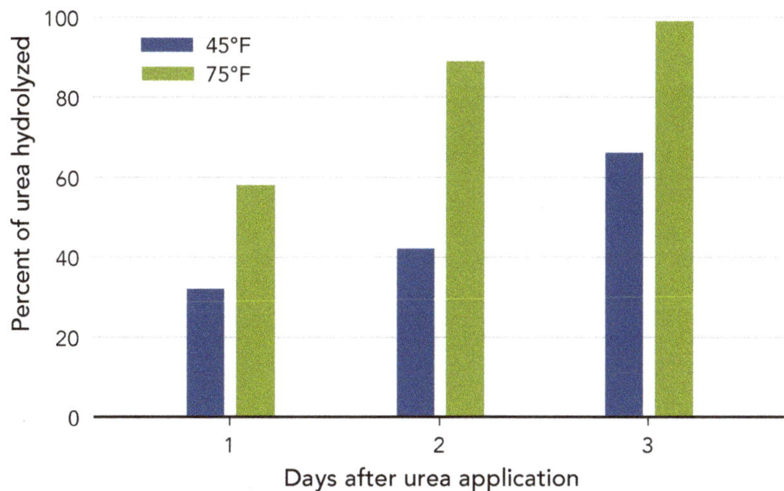

FIGURE 4.7.

Effect of soil temperature on the speed of urea hydrolysis; columns represent the mean of four California soils.
Source: Adapted from Broadbent et al. 1958.

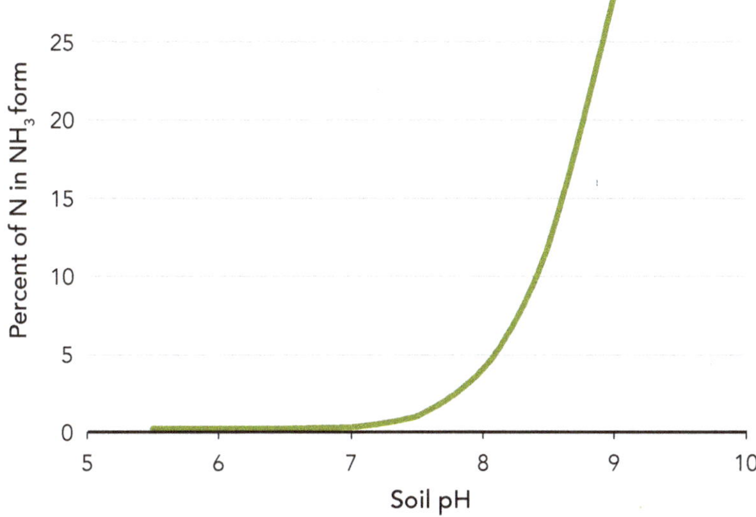

FIGURE 4.8.

Effect of pH on the equilibrium between ammonia (NH₃) and ammonium (NH₄⁺) in soil solution or irrigation water.

drous ammonia is applied to moist soil or runs into irrigation water, a chemical reaction forms NH_4^+ and hydroxide (OH^-):

$$NH_3 + H_2O \Longrightarrow NH_4^+ + OH^-$$

Hydroxide dramatically increases pH; for example, running anhydrous ammonia into irrigation water can raise the pH of the water above 9.0, creating an opportunity for large volatilization losses. Urea hydrolysis also produces alkalinity, raising the pH around the fertilizer prills. When urea is applied below the soil surface, volatilization is minimal. However, when urea is broadcast without incorporation and no irrigation or rainfall occurs to move it into the soil, the pH around the prills can increase enough to allow substantial volatilization to occur.

NITROGEN FERTILIZERS

A wide range of N-containing synthetic fertilizers are used in California vegetable production (table 4.1). Most blended N/P/K fertilizers are made using these constituents. Fertilizers containing phosphorus are predominately applied preplant, while in-season N application is dominated by the liquid fertilizers urea ammonium nitrate (UAN-32), ammonium nitrate (AN-20), and calcium ammonium nitrate (CAN-17). A number of considerations are involved in choosing which N fertilizer fits a particular situation: cost, physical state (solid or liquid), solubility, N form (urea, NH_4^+ or NO_3^-), companion ions (K^+, Ca^{2+}, SO_4^{2-}, etc.), toxicity to seedlings, and so on. Given the relatively rapid rate of nitrification in California soils, the form of N applied (urea, NH_4^+, or NO_3^-) generally has little bearing on agronomic performance because the vast majority of N is in nitrate form before crop uptake. A possible exception is during winter production, when low soil temperature slows nitrification and prolongs the presence of NH_4^+. Under such circumstances, some growers favor the use of nitrate fertilizers.

Nitrogen applied in nitrate form has no impact on soil pH, but N applied as urea or NH_4^+ acidifies soil as the N is nitrified (H^+ ions are released in the nitrification reaction). Soil acidification is of little consequence in calcareous, highly buffered soils; in such soils, ammonium fertilizers can be used for many years with only minimal effect on soil pH. In soils lacking free lime, acidification from N fertilizer use may be observed over just a few years.

Fertilizer Technology to Improve Nitrogen Efficiency

A range of technologies has been developed to improve the field efficiency of N fertilizers. These technologies encompass three basic

TABLE 4.1.

Common N fertilizers used in California vegetable production

Fertilizer	State	Typical analysis N-P_2O_5-K_2O	Soil acidification potential*
ammonium nitrate (AN-20)	liquid	20-0-0	3.6
ammonium polyphosphate	liquid	10-34-0	7.2
ammonium sulfate	dry	21-0-0	7.2
ammonium thiosulfate	liquid	12-0-0-26 S	10.8
calcium ammonium nitrate (CAN-17)	liquid	17-0-0-9 Ca	< 2.0
monoammonium phosphate (MAP)	dry	11-52-0	7.2
potassium nitrate	dry	13-0-46	—
urea	dry	46-0-0	3.6
urea ammonium nitrate (UAN-32)	liquid	32-0-0	3.6

*Pounds of calcium carbonate equivalent per pound of N needed to neutralize acidity produced during nitrification.

modes of action: manipulating soil N transformations, modifying the timing of nutrient release, and stimulating plant nutrient uptake.

Urease Inhibition

Nitrogen losses through ammonia volatilization can be significant from urea-containing fertilizers (granular urea and liquid blends containing urea). N-(n-butyl) thiophosphoric triamide (NBPT) is a urease inhibitor sometimes added to urea-containing fertilizers to slow hydrolysis and reduce possible ammonia volatilization. Fertilizer application methods and environmental conditions determine the relative risk of volatilization. Volatilization is an issue primarily when urea is broadcast and not quickly incorporated with rain or irrigation; under such conditions, NBPT can be very effective. However, in vegetable production, urea-containing fertilizers are usually banded below the soil surface or applied with water (as with UAN-32); with these practices, the potential for volatilization is very limited, and NBPT would have minimal effect.

Nitrification Inhibition

One approach to limiting the nitrate leaching potential would be to slow the rate of nitrification of ammonium fertilizers; NH_4^+ is much less mobile than NO_3^- in soil. Nitrification inhibitors have been available for several decades, but their use in vegetable production has been limited. Nitrification inhibitors suppress the activity of nitrifying bacteria and therefore are, with the exception of dicyandiamide (DCD), classified as pesticides subject to regulation by the California Department of Pesticide Regulation (DPR). Until recently no proprietary nitrification inhibitors were registered by DPR for use in California vegetable production.

Aside from the registration issue, the utility of nitrification inhibitors in vegetable production is limited. These products are best suited for field conditions in which there is a risk of substantial nitrate loss through leaching and in which N application is made far ahead of crop N uptake; a prime example would be preplant N fertilization of corn in the Midwest. However, in California vegetable production nitrate leaching due to rain is a minor issue (except for winter production) because irrigation can be controlled to minimize leaching (particularly with drip irrigation) and multiple N applications throughout the season are a common practice. Extensive field trials in coastal leafy greens production have shown limited improvement in N uptake efficiency with the use of either DCD or proprietary nitrification inhibitors (Smith et al. 2016a).

Slow Release/Controlled Release Fertilizer

Several approaches other than inhibition of nitrification have been developed to control the rate of N release from fertilizers, limiting the amount of N at risk of leaching at any one time. One approach has been to develop complex chemical N forms that require microbial action to make the N plant available. These products are generally referred to as slow release fertilizers (SRFs). They are made by polymerizing urea molecules in various ways; the most common types of SRF are methylene urea, urea formaldehyde, and triazone polymers. Since N release from SRFs requires microbial action, the rate of N release is strongly affected by soil temperature.

The other main category of products is referred to as controlled release fertilizers (CRFs). These fertilizers contain prills of soluble fertilizer (either urea or N/P/K blends) surrounded by a porous plastic coating. Nutrient release from CRFs is dependent on diffusion through the coating, not microbial activity; the rate of diffusion is controlled by the chemical composition and thickness of the polymer coating. Soil temperature has a smaller effect on the process of diffusion than on microbial activity, so the rate of nutrient release from CRFs is less affected by temperature than that from SRFs. CRFs are rated for the speed of nutrient release; for example, when a fertilizer is labelled as a 60-day product it means that approximately 80 percent of its nutrient content is expected to be released within 60 days at a moderate soil temperature (around 70°F).

While the potential advantage of SRFs or CRFs in limiting nitrate leaching loss is clear, whether any measureable advantage in N uptake efficiency is achieved depends on field conditions and grower management. In general, these fertilizers are most useful where in-season nitrate leaching potential is significant but beyond the control of the grower. The classic example is overwinter production, where rain can leach the soil profile and wet field conditions can limit timely N application. The production of baby greens (spinach, lettuce, etc.) presents similar challenges, in that a fast-growing, shallow-rooted crop is grown from seed using sprinkler irrigation; a substantial amount of leaching is nearly inevitable in this situation. However, for summer production of vegetable crops that are reasonably deep rooted, or where drip irrigation can

be used, a grower has a substantial degree of control over in-season NO_3-N leaching. In such circumstances, SRFs or CRFs offer little benefit.

Even where a case can be made for the use of CRFs, there are challenges to achieving an agronomic or environmental benefit. Field trials conducted under coastal California conditions have shown that CRFs seldom produce measureable benefits compared with well-managed conventional fertilizers (Hartz and Smith 2009). A fundamental problem is that the nutrient release rate from a CRF is relatively linear after soil application, while crop N uptake is sigmoidal (see fig. 2.2). Preplant CRF application runs the risk of substantial N release before the crop can use it; sidedressing a CRF during crop growth runs the risk that the crop N uptake requirement will exceed the fertilizer N release rate. As the vegetable industry continues to increase the use of drip irrigation, which improves irrigation control and expands the opportunity to fertigate N on demand, CRFs will likely remain a niche product.

Humic Substances

Products derived from leonardite and lignite, sedimentary materials similar to coal, have been added to fertilizer for decades. The assumed active ingredients in such products are humic and fulvic acids; for this discussion, these products are referred to as humates. Among the beneficial properties that have been attributed to humates are improved soil structure, stimulation of soil microbial activity, enhancement of root growth, and chelation of nutrients. Many lab and greenhouse studies have confirmed some of these claims, with far fewer studies documenting positive effects from the use of humates in the field.

Experts differ as to whether humates have consistently positive effects under representative field conditions. Canellas et al. (2015) suggested that the use of humates often elicits a positive crop response. Conversely, Chen et al. (2004) concluded that field application, at typical industry rates, is unlikely to significantly affect crop performance. Chen based this conclusion on the fact that soil contains soluble organic compounds chemically

TABLE 4.2.

Comparison of crop N uptake with seasonal N fertilization rate in commercial vegetable fields

Crop	Number of fields monitored	N fertilizer rate	Crop N uptake	Reference
		Average lb N/acre		
broccoli	14	186	327	Smith et al. 2016b
cabbage	8	249	327	Smith et al. 2016b
cauliflower	8	228	284	Smith et al. 2016b
processing tomato	14	191	274	Lazcano et al. 2015

similar to those in humates; these soil compounds undoubtedly perform some of the same biostimulatory functions of humates. Except in soils with very low organic matter, the concentration of these soluble organic compounds naturally present in soil is usually adequate to maximize any crop response.

There is limited California field data on vegetable crop response to humate products. Hartz and Bottoms (2010) evaluated five commercial humate products for their effects on soil microbial activity and plant growth. In lab and greenhouse experiments humates measurably affected soil microbial activity but had minimal effect on lettuce growth or nutrient uptake. In two field experiments, the application of humates with preplant fertilizer had no effect on tomato growth, nutrient uptake, or yield.

NITROGEN RATE AND TIMING

Nitrogen Rate

It is inappropriate to give generic N rate recommendations for California vegetable crops because conditions vary so greatly across fields, production areas, and seasons. Developing an efficient N management program requires consideration of field-specific conditions. A starting point in developing an efficient N management plan is to understand the N uptake requirement of the crop. Table 2.1 lists typical crop N uptake rates; the high end of the range represents vigorous fields grown under summer conditions, while the low end of the range represents fields grown in cool weather or fields with low yield potential. Plant population can also be a factor; leafy green crops have higher plant populations when grown on 80-inch beds than on 40-inch beds, and N uptake is therefore usually higher on 80-inch beds. The values in table 2.1 represent the N uptake observed in commercial fields and undoubtedly include some luxury uptake.

As table 4.2 demonstrates, the entire crop N uptake requirement does not have to be supplied from N fertilizer. Across a range of vegetables it is clear that nonfertilizer sources of N contribute substantially to crop fertility; in broccoli, N uptake often exceeds the N fertilization rate by more than 100 pounds per acre. Even for lettuce, a relatively shallow-rooted crop widely believed to be sensitive to N deficiency, Bottoms et al. (2012) documented that with careful management high yields could be produced with seasonal N rates substantially below the rate of crop N uptake.

Important nonfertilizer sources of N are residual soil NO_3-N, in-season soil N mineralization, and NO_3-N in irrigation water. Integration of these N sources into N fertilizer rate decisions is a complex matter, and no firm rules apply to all field situations. A discussion of soil N mineralization was given earlier in this chapter, and guidelines for crediting residual soil NO_3-N and irrigation water N are covered in chapters 9 and 10, respectively.

In certain circumstances, the seasonal N fertilization rate may need to exceed the expected crop N uptake. However, for most crops and most field conditions, careful management makes it possible to produce a good crop with a seasonal N fertilization rate at or lower than the crop N uptake. This is

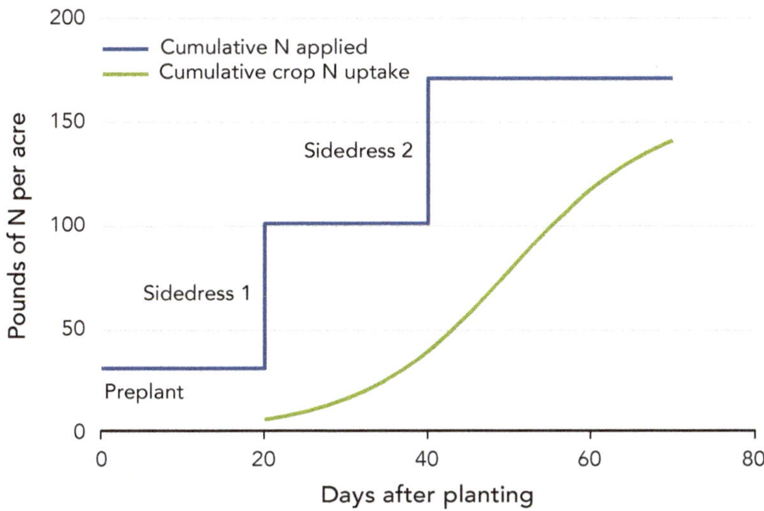

FIGURE 4.9.

Comparison of N application timing with crop N uptake in a traditional lettuce fertilization program.

obviously easier for deep-rooted crops (carrot, broccoli, melon, tomato, etc.) than for more shallowly rooted crops such as lettuce and spinach. Crop rotation and time of year also have important impacts. Summer lettuce grown following a spring broccoli crop is likely to benefit from a substantial amount of N mineralized from the broccoli residue. By contrast, spring lettuce grown after a wet winter may encounter virtually no residual soil NO_3-N.

Spinach and other baby greens present a particular challenge, in that they are started from seed using sprinkler irrigation, are shallow rooted, and have a high daily N uptake rate due to their high plant density. These crops generally require higher N fertilization rates than crop N uptake to produce high yield and high quality. Ideally, these baby greens should be rotated with more deeply rooted crops that could scavenge nitrate from deeper in the soil, limiting N loss to the environment.

Given the high value of vegetable crops relative to the cost of N fertilizer, there has historically been little economic incentive to strive for maximum N efficiency. However, increasing regulatory pressure on growers to reduce N loss to the environment now provides a powerful incentive to improve N management.

Nitrogen Timing

Historically, a typical N fertilization plan included a preplant application (usually N and P) and one or more N sidedressings. In this system, N application occurred substantially in advance of crop N uptake (fig. 4.9). This "front loading" meant that soil nitrate concentration was high through much of the season, maximizing the opportunity to leach nitrate with irrigation. A system in which N application more closely matched the timing of crop N uptake would be more efficient.

Preplant application is generally the least efficient timing for N fertilization. Not all fields require preplant N fertilization. Due to California's semiarid climate and the high intensity of production, a substantial amount of residual soil nitrate is often present in vegetable fields. This is particularly true of coastal production areas where multiple crops per year are grown; the need for preplant N application in summer-planted fields following spring vegetables is typically minimal. Depending on the amount of winter rain and preplant irrigation, substantial residual soil nitrate may also be present in Central Valley fields. Preplant P management affects preplant N rate because all common P fertilizers contain N as well. Where preplant P application is needed, using a fertilizer with a high P:N ratio limits the amount of N applied. It should be noted that not all fields require P fertilization. Past fertilization practices, particularly in coastal regions, have elevated soil P availability to the point that many fields are unresponsive to P fertilization; see chapter 5 for details.

Ideally, in-season N application would be managed to stay just ahead of crop N uptake. In drip-irrigated fields, one could theoretically "spoon-feed" the crop, applying N with each irrigation; in practice, as long as irrigation is managed to limit the leaching volume, little is gained by fertigating more often than every 7 to 14 days. Traditional sidedressing remains the primary N application method in furrow- or sprinkler-irrigated fields, which imposes some constraints; applying more than two sidedressings is unusual. Delaying or reducing the first sidedressing where substantial residual soil nitrate is present can dramatically improve N efficiency (see chapter 9). Since the distribution uniformity (DU) of furrow and sprinkler irrigation is often poor, applying N in irrigation water is not an efficient practice; in general, the DU of water-run fertilizer is no better, and often worse, than the DU of water application. The importance of irrigation efficiency on N use efficiency is discussed in chapter 10.

NITROGEN EFFECTS ON PRODUCT QUALITY AND POSTHARVEST LIFE

In addition to the effects of N management on crop yield, some growers are convinced that N management can also substantially influence crop quality. That is certainly true in the sense that an insufficient N supply can affect the color of leafy greens and the size of heads (lettuce, cauliflower) or fruit (melon, pepper). However, N application in excess of that needed to maximize crop yield is unlikely to improve product quality or postharvest life. Hartz et al. (2000) and Breschini and Hartz (2002) conducted trials evaluating the need for N sidedressing in commercial lettuce fields with high residual soil NO_3-N. In these trials, reducing the seasonal N application rate by skipping the first sidedressing did not affect crop yield and did not affect either the color or the postharvest life of harvested heads. In fact, in certain circumstances excessive N fertilization can reduce quality. Cracking of carrots, a major quality defect, increased with N application in excess of that required to maximize root yield (Hartz et al. 2005).

REFERENCES

Agehara, S., and D. D. Warncke. 2005. Soil moisture and temperature effects on nitrogen release from organic nitrogen sources. SoilScience Society of America Journal 69:1844–1855.

Bottoms, T. G., R. F. Smith, M. D. Cahn, and T. K. Hartz. 2012. Nitrogen requirements and N status determination of lettuce. HortScience 47:1768–1774.

Breschini, S. J., and T. K. Hartz. 2002. Presidedress soil nitrate testing reduces nitrogen fertilizer use and nitrate leaching hazard in lettuce production. HortScience 37:1061–1064.

Broadbent, F. E., G. N. Hill, and K. B. Tyler. 1958. Transformations and movement of urea in soils. Soil Science Society of America Proceedings 22:303–307.

Canellas, L. P., F. L. Olivares, N. O. Aguiar, D. L. Jones, A. Nebbioso, P. Mazzei, and A. Piccolo. 2015. Humic and fulvic acids as biostimulants in horticulture. Scientia Horticuturae 196:15–27.

Castro Bustamante, S., and T. K. Hartz. 2016. Carbon mineralization and water-extractable organic carbon and nitrogen as predictors of soil health and nitrogen mineralization potential. Communications in Soil Science and Plant Analysis 47 Supplement 1:46–53.

Chen, Y., M. De Nobili, and T. Aviad. 2004. Stimulatory effects of humic substances on plant growth. In F. Magdoff and R. R. Weil, eds., Soil organic matter in sustainable agriculture. Boca Raton: CRC Press.

Hartz, T. K., and T. G. Bottoms. 2010. Humic substances generally ineffective in improving vegetable crop nutrient uptake or productivity. HortScience 45:906–910.

Hartz, T. K., and R. F. Smith. 2009. Controlled release fertilizer for vegetable production: The California experience. HortTechnology 19:20–22.

Hartz, T. K., W. E. Bendixen, and L. Wierdsma. 2000. Pre-sidedress soil nitrate testing as a nitrogen management tool in irrigated vegetable production. HortScience 35:651–656.

Hartz, T. K., P. R. Johnstone, and J. J. Nunez. 2005. Production environment and nitrogen fertility affect carrot cracking. HortScience 40:611–615.

Havlin, J. L., S. L. Tisdale, W. L. Nelson, and J. D. Beaton. 2014. Soil fertility and fertilizers. Upper Saddle River, NJ: Pearson.

Krueskopf, H. H., J. P. Mitchell, T. K. Hartz, D. M. May, E. M. Miyao, and M. D. Cahn. 2002. Pre-sidedress soil nitrate testing identifies processing tomato fields not requiring sidedress N fertilizer. HortScience 37:520–524.

Kumar, K. M., and K. Goh. 2000. Crop residues and management practices: Effects on soil quality, soil nitrogen dynamics, crop yield and nitrogen recovery. Advances in Agronomy 68:197–319.

Lazcano, C., J. Wade, W. R. Horwath, and M. Burger. 2015. Soil sampling protocol reliably estimates preplant NO_3 in SDI tomatoes. California Agriculture 69:222–229.

Mitchell, J., T. Hartz, S. Pettygrove, D. Munk, D. May, F. Menezes, J. Diener, and T. O'Neill. 1999. Organic matter recycling varies with crops grown. California Agriculture 53(4): 37–40.

Ryden, J. C., and L. J. Lund. 1980. Nature and extent of directly measured denitrification losses from some irrigated vegetable crop production units. Soil Science Society of America Journal 44:505–511.

Smith, R., M. Cahn, and T. K. Hartz. 2016a. Evaluation of N uptake and water use of leafy greens grown in high-density 80-inch bed plantings, and demonstration of best management practices. CDFA Fertilizer Research and Education Program Final Report 12-0362-SA. CDFA website, www.cdfa.ca.gov/is/ffldrs/frep/pdfs/completedprojects/12-0362-SA_Smith.pdf

Smith, R., M. Cahn, T. K. Hartz, P. Love, and B. Farrara. 2016b. Nitrogen dynamics of cole crop production: Implications for fertility management and environmental protection. HortScience 51:1586–1591.

5

Phosphorus Management

Phosphorus availability in soils is a complex topic. Luckily, mastering only a few basic concepts allows one to effectively manage P fertilization. In the soil, P exists in a number of different forms, or pools (fig. 5.1). These P pools exist in an equilibrium. In the relatively low-organic-matter soils common in California, P mineralization from soil organic matter is usually a minor factor in soil P availability. The important interactions are between the inorganic pools: soil solution (soluble) P, P adsorbed on the surfaces of soil minerals, and P contained in secondary minerals. When fertilizer P is added, soil solution P temporarily increases, but over days to months it declines as P becomes adsorbed on clay or mineral surfaces or precipitates in secondary minerals of low solubility. Conversely, as plants remove P from soil solution, the solution is replenished from the other pools.

The Phosphorus Cycle

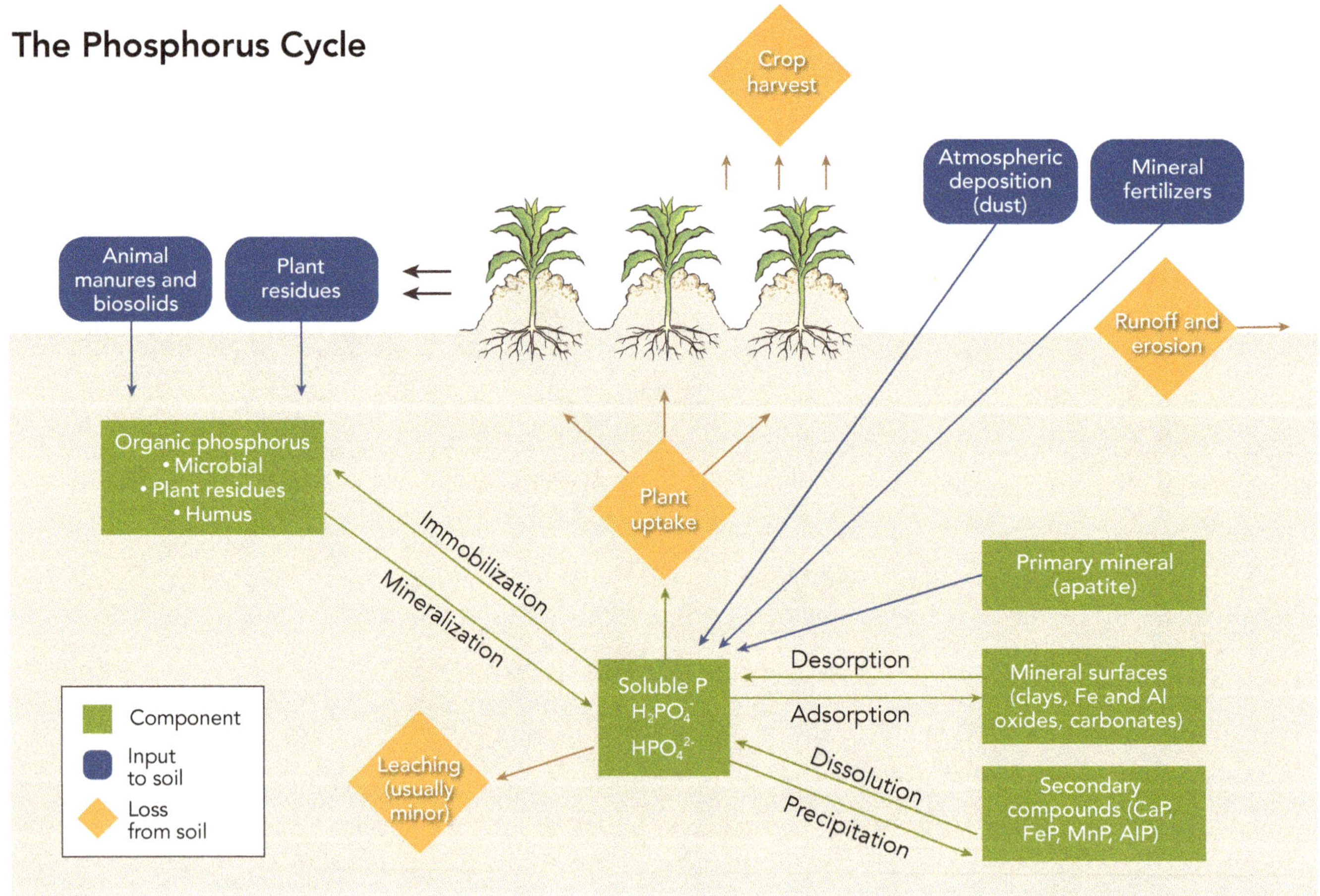

FIGURE 5.1.

The agricultural phosphorus cycle.
Source: Redrawn from diagram by Wikipedia user Welcome1To1The1Jungle per Creative Commons license CC-BY-3.0.

The importance of this equilibrium between the inorganic P pools can be illustrated by comparing the amount of P in soil solution with the P uptake requirement of vegetable crops. Soil solution P is well below 1.0 ppm in most agricultural soils. A loam soil at field capacity moisture content would therefore have less than 0.4 pounds of elemental P in the soil solution per acre-foot of soil. The daily P uptake of vegetable crops can exceed 0.7 pounds of P per acre per day, almost all of which is taken up from soil solution. Unless soil solution P concentration can be replenished quickly, crop growth would suffer.

The importance of diffusive movement of P in plant P uptake can be illustrated using a similar example. An acre of tomatoes at full bloom growth stage may take up 0.7 pounds of P daily while transpiring 60,000 pounds (about 7,200 gallons) of water. If the soil solution P concentration were 1.0 ppm, the transpired water would contain only 0.06 pounds of P, or less than 10 percent of daily P uptake. Clearly, mass flow (P movement toward a root in transpirational flow) is a minor factor in P supply. The only way plants can get enough P from the soil solution is by diffusion (P movement from areas of high concentration toward lower concentration). As a root removes P from soil solution, additional P diffuses toward the root. Since the effective diffusive

distance for P in soil is very limited (typically less than 1 millimeter per day), roots are able to access soil P only in their immediate vicinity. This limitation on the distance that P can move in soil is a major reason why applying P in a band close to a developing seedling is more effective than a broadcast application. The limited diffusive movement of P is also the reason why placement of P relative to a transplant or a developing seedling is so important.

EVALUATING SOIL PHOSPHORUS AVAILABILITY

The common soil P extraction tests (Olsen and Bray, see chapter 3) extract all solution P, much of the adsorbed P, and some of the most soluble forms of mineral P. While the P extracted represents only a small fraction of total soil P (usually from 2 to 10 percent), it is the fraction most accessible to crops. Some authorities have suggested that a simple chemical extraction test cannot accurately estimate soil P availability, given the complex nature of the equilibrium among soil P pools and other confounding effects (soil pH, presence of free lime, etc.). Various dynamic test procedures designed to mimic the action of plant roots have been proposed, including incubating anion exchange resins in moist soil; some commercial labs offer such tests. The anion resin attracts $H_2PO_4^-$ or HPO_4^{2-} ions, removing them from solution like plant roots do. The rate at which the resin accumulates these ions is a measure of how readily soil solution P is replenished from the other soil P pools. Research results have been mixed, but in general reasonably good correlation has been found between resin tests and standard extraction tests such as Olsen (fig. 5.2).

Standard soil P tests provide a good picture of the P equilibrium in soils. Figure 5.3A shows the correlation between Olsen P and soluble soil P (P extractable in a weak solution of calcium chloride, a surrogate measurement for soil solution P). As Olsen P increases, soluble soil P increases. Figure 5.3B shows the relationship between Olsen P and the degree of soil P saturation (the point at which a soil

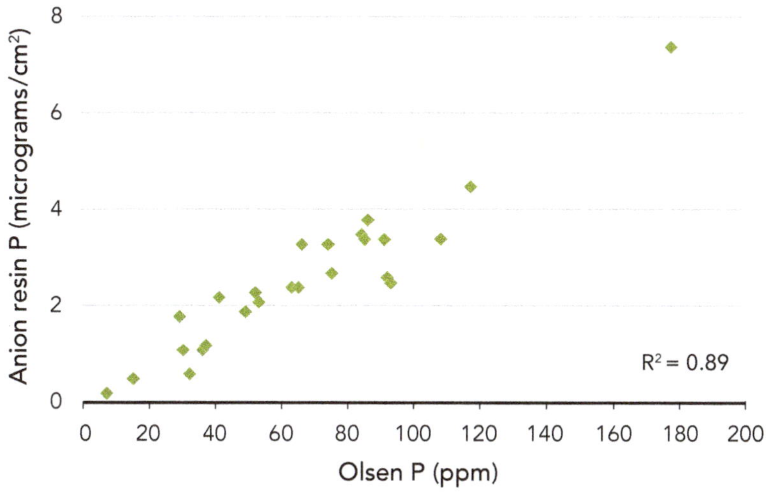

FIGURE 5.2.

Correlation between the Olsen soil P test and P capture by anion resin for twenty-five Salinas Valley soils in vegetable crop rotations. Source: Johnstone et al. 2005.

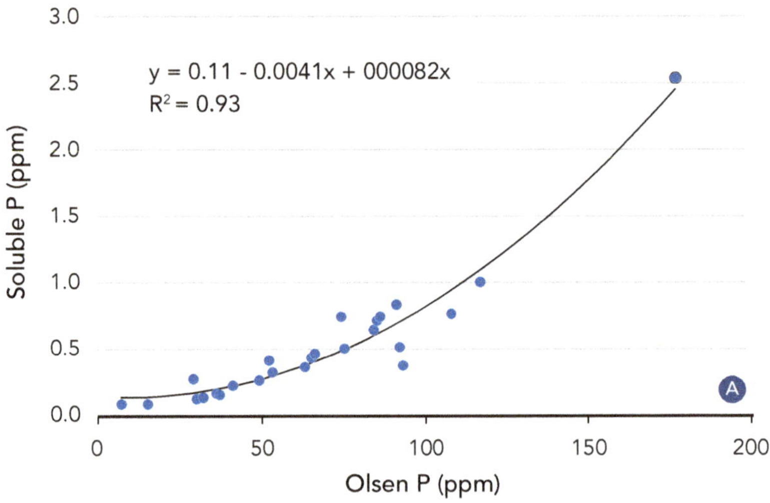

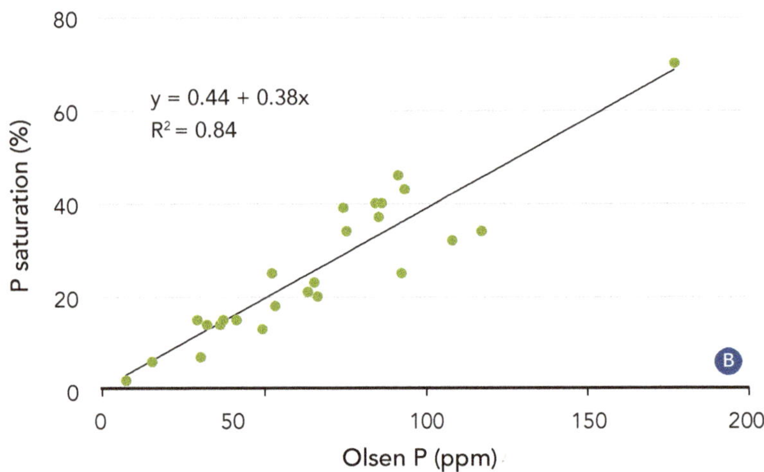

FIGURE 5.3.

The relationship between Olsen P and soluble soil P (a) and soil P saturation (b) for twenty-five Salinas Valley soils.
Source: Adapted from Hartz and Johnstone 2006.

P lost by these routes is quite small, although still potentially environmentally damaging. Consequently, most of the P applied but not removed in harvested products remains in the soil. This causes an increase in soil test P over time, but that increase is slow because much of the residual P is fixed (adsorbed or precipitated). An increase in soil test P over time shows that the equilibrium between the soil P pools is shifting toward higher P availability.

The long-term effect of P fertilization practices can be shown by comparing production practices in the coastal regions with those of the Central Valley. In coastal areas from Ventura to Watsonville, most fields have been in vegetable rotations for many years, and production of multiple crops per year has been the norm. A grower applying P fertilization at typical rates has been increasing soil P content substantially with each crop produced. By contrast, a Central Valley grower producing processing tomatoes may be modestly enriching soil P with each tomato crop, but only one crop per year is grown, and tomatoes have historically been rotated with lower-value row crops, which are more lightly fertilized. These different P management histories have resulted in stark regional differences in soil P availability. Central Valley soils are generally below 30 ppm Olsen P, not substantially different from their original state before irrigated farming began. Many coastal vegetable soils have greater than 50 ppm Olsen P, with a substantial percentage of fields approaching or exceeding 100 ppm. Obviously, the fertilization history of individual fields may differ from these regional trends; for example, some Central Valley fields have elevated Olsen P resulting from heavy P fertilization or repeated manure applications.

The important point about historical P management is that as soil test P increases over time, traditional P fertilization rates can be reduced without endangering crop productivity. Some growers continue to use standard P fertilization programs developed many years ago, when soils were highly deficient in P; today, with much higher soil P availability, such practices waste money and increase the risk of surface water pollution (see chapter 12).

cannot fix any additional P through adsorption or precipitation). As Olsen P increases, the soil's ability to fix P decreases. The bottom line is that the Olsen test provides a reliable estimate of relative soil P availability and indicates the state of the equilibrium among the soil P pools.

Phosphorus fertilization practices can affect the soil P equilibrium over time. In the context of vegetable production, P fertilization rates are usually higher than the amount of P removed with harvested products (table 5.1). Unlike N, which can be lost in large quantities through leaching and runoff, the amount of

TABLE 5.1.

Comparison of current P fertilization rates with P harvest removal

Crop	Typical range (lb P_2O_5/acre)		P balance (lb P_2O_5/acre)
	Application	Harvest removal	
broccoli	60–90*	25–35	35–55
lettuce	60–90*	20–25	40–65
processing tomato	50–100	45–60	5–40

*Values represent typical application rates in coastal production; P rates in the Imperial Valley are generally higher.

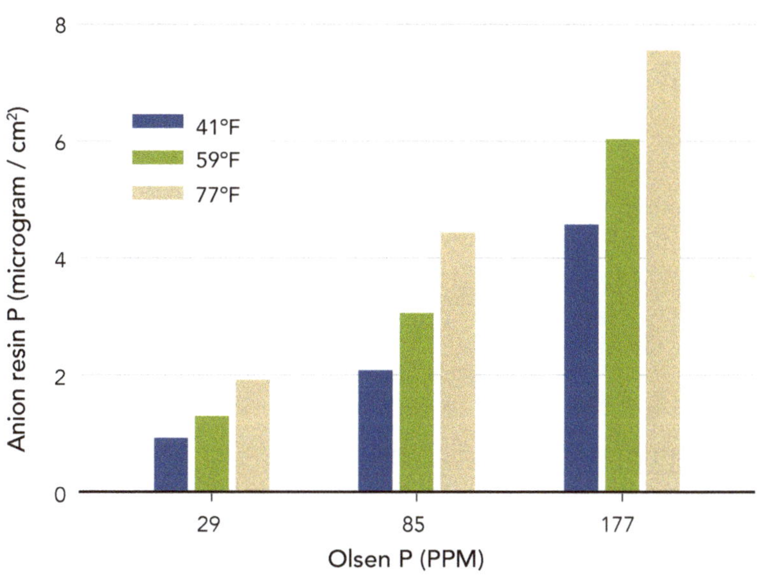

FIGURE 5.4.

Effect of soil temperature on anion resin P capture in three soils of differing Olsen P concentration.
Source: Johnstone et al. 2005.

FACTORS AFFECTING SOIL PHOSPHORUS AVAILABILITY

Mycorrhizae

Mycorrhizae are soil fungi that form symbiotic associations with plant roots. Among the many potential benefits attributed to mycorrhizae is enhanced soil P availability. Mycorrhizae are widely distributed in agricultural soils, and colonization of roots is a common phenomenon for many vegetable crops. Some crops (e.g., citrus) are particularly dependent on mycorrhizal associations and may grow poorly in soils lacking compatible mycorrhizae, such as fumigated soils. However, in vegetable crop production mycorrhizal association generally has consistent value only in soils with very limited P availability. In the vast majority of California vegetable fields, root colonization by mycorrhizae makes no substantive difference to P nutrition. Mycorrhizal association with roots may have other beneficial effects, including improving micronutrient availability and soil aggregate stability (Cavagnaro et al. 2006).

Soil Temperature

Soil temperature affects soil P availability. The microbial and chemical processes that control P availability are slowed as temperature decreases. Also, at lower soil temperature roots grow more slowly, limiting the volume of soil explored. To quantify soil temperature effects on soil P availability, the same anion resin incubation technique referred to in fig. 5.2 was used to compare the amount of P adsorbed from soils at three different temperatures (fig. 5.4). Temperature effects were similar in all soils, with adsorbed P increasing approximately 2 to 3 percent for every 1°F increase in soil temperature. This effect is clearly significant enough to justify adjusting P fertilization practices to account for soil temperature.

Soil pH

The P fixation potential of a soil is affected by its pH and the presence of Al, Fe, and Ca. At soil pH below about 6.0, P can be adsorbed onto Al or Fe oxides or precipitated as Al or Fe minerals. At higher pH, P can be adsorbed on $CaCO_3$ (free lime) or precipitated as Ca minerals. Since California soils are overwhelmingly above pH 6.0, soil Ca status usually controls P fixation. Soils rich in free lime tend to fix a higher percentage of applied

TABLE 5.2.

Soil solution P concentration associated with maximum growth of selected crops

Crop	Soil solution P (ppm)
wheat	0.04
corn	0.05
sweet potato	0.10
potato	0.18
cabbage	0.20
tomato	0.20
lettuce	0.30

Source: Hue and Fox 2010.

P, meaning that higher P fertilization rates may be needed to achieve the desired increase in soil P availability. Although it has not been independently verified in California trials, some authorities suggest increasing P application by as much as 10 pounds P_2O_5 per acre for each percent of free lime in the soil (Leytem and Mikkelsen 2005).

PHOSPHORUS FERTILIZERS

Mineral P fertilizers are manufactured by treating mined phosphate rock, in which the phosphorus is not directly plant available, with acid to produce materials in which most or all of the P is in a soluble form available for plant uptake. Examples of common P fertilizers are

- triple superphosphate (0-45-0)
- nitrophosphate (analyses vary widely)
- ammonium polyphosphate (10-34-0)
- monoammonium phosphate (11-52-0)
- diammonium phosphate (18-46-0)
- phosphoric acid (white form, 0-54-0; green form, 0-52-0)

The P content of most of these fertilizers is overwhelmingly in the orthophosphate (PO_4-P) form, which is soluble and readily available for plant uptake. The exception is ammonium polyphosphate, which contains a mixture of orthophosphate and polyphosphate (polymer phosphate chains requiring the action of soil enzymes to break down into PO_4-P). This enzymatic breakdown occurs relatively quickly in warm, moist soil and is usually complete within several weeks after application. The agronomic performance per unit of P content is similar with all of these products. What product is most appropriate depends on a number of factors, such as cost, form (liquid or dry), compatibility with equipment, ratio of P to N, salt index, safety to developing seedlings, etc.

PHOSPHORUS RATE, TIMING, AND APPLICATION TECHNIQUE

Crops have different P uptake requirements and remove different amounts of P in harvested products (see table 2.1). Crops also require different levels of P in soil solution to achieve maximum growth (table 5.2). Grain crops such as corn or wheat are highly efficient in taking up P and therefore can thrive in relatively low-P soil. Most vegetable crops require much higher soil solution P for peak production; there are also substantial differences among vegetable crops. Since soil solution P is strongly correlated with soil test P, soil test thresholds for response to P fertilization therefore differ among vegetable crops (see table 3.5).

Given the variety of factors that can affect soil P availability, it is clear that P fertilization rates should be determined based on field-specific conditions. However, some general guidelines can be used.

- If soil test P is well below the crop response threshold, the P fertilizer rate should exceed the total crop P uptake rate. In extreme cases, the appropriate rate may be several times the crop P uptake.
- If soil test P is close to the crop response threshold, the P fertilizer rate should be close to the expected rate of harvest P removal.
- If soil test P is well above the crop response threshold, no P fertilizer is likely to be required, but if a P application is made it should be limited to no more than the crop P harvest removal to limit additional soil P enrichment.

As an example, consider the production of romaine lettuce. This crop takes up approximately 14 pounds of elemental P per acre (32 pounds P_2O_5 per acre) and removes about 60 percent of that in the harvested product. Romaine has a high Olsen P response threshold (response possible up to 40 to 60 ppm under cool soil temperature). Therefore, a summer field in Santa Maria with an Olsen P value of 50 ppm may be unresponsive to P fertilization, and P application could safely be limited to 30 pounds P_2O_5 per acre or less. Conversely, in a winter romaine field in the Imperial Valley with an Olsen value of 20 ppm and a considerable amount of free lime, a P fertilization rate of more than 100 pounds P_2O_5 per acre could be justified.

Application of P either preplant or at planting is the most efficient approach in most circumstances. This is because crops are most likely to encounter limited soil P availability early in the crop cycle. For spring-planted crops, soil temperature is lowest during crop establishment, which reduces soil P availability. Also, seedlings have a very limited root system and can access P from only a small fraction of the soil volume. As soil warms and the root system expands, the crop's ability to extract P from the soil increases. If preplant or at-planting P fertilization is managed appropriately, in-season P application is seldom required.

Phosphorus should be applied as close to the planting date as practical because applied P tends to become less available over time, converting from soluble P into less-available P forms. Within 2 months of application, up to half of fertilizer P may be fixed and no longer extractable by standard soil tests. The potential for soil P fixation varies among soils with abnormally high or low pH, high clay content, or a high level of free lime increasing fixation. Fall P application for spring planting is not an efficient practice.

The placement of fertilizer P can be important. Fertilizer P does not move freely in soil, in general remaining within several inches of the site of application. This can be either a benefit or a limitation, depending on the field situation. Limited P movement can be a benefit: when applied in a concentrated band, the zone around the band retains high P availability for an extended period because it limits contact with soil minerals that could fix P. If the band is placed appropriately, it maximizes P availability to developing seedlings at that critical period of development.

In some circumstances limited P movement may be detrimental to crop productivity. In soil of very low P availability (Olsen P below 10 ppm), a band of preplant P fertilizer may improve P accessibility at the seedling stage, but the limited soil volume affected by the band may be inadequate to supply the high rate of P uptake required during the rapid growth phase. Another example is the application of P through buried drip irrigation, as is commonly done in tomato production. The fertigated P may not move close enough to the transplant root ball to affect early growth.

Efficient P management requires balancing a crop's P requirement against the soil's P availability, which can be a complex task. Many growers use a standard P fertilizer program, with application rates high enough to maximize productivity in all field circumstances. This approach simplifies management and ensures P sufficiency. However, applying the same P rate in fields of low and high soil P availability is not only economically inefficient, but can also be environmentally damaging (see chapter 12).

REFERENCES

Cavagnaro, T., L. E. Jackson, J. Six, H. Ferris, S. Goyal, D. Asami, and K. M. Scow. 2006. Arbuscular mycorrhizas, microbial communities, nutrient availability, and soil aggregates in organic tomato production. Plant and Soil 282:209–225.

Hartz, T. K., and P. R. Johnstone. 2006. Relationship between soil phosphorus availability and phosphorus loss potential in runoff and drainage. Communications in Soil Science and Plant Analysis 37:1525–1536.

Hue, N. V., and R. L. Fox. 2010. Predicting plant phosphorus requirements for Hawaii soils using a combination of phosphorus sorption isotherms and chemical extraction methods. Communications in Soil Science and Plant Analysis 41:133–143.

Johnstone, P. R., T. K. Hartz, M. D. Cahn, and M. R. Johnstone. 2005. Lettuce response to phosphorus fertilization in high phosphorus soils. HortScience 40:1499–1503.

Leytem, A. B., and R. L. Mikkelsen. 2005. The nature of phosphorus in calcareous soils. Better Crops 89(2): 11–13.

Potassium Management

Potassium is the macronutrient that vegetable crops take up in the greatest quantity; K uptake by vegetable crops is often 10 to 20 percent greater than N uptake. Potassium is also the only macronutrient for which the rate of removal in the harvested product often exceeds the fertilizer application rate. Instead of enriching the soil over time, traditional fertilization practices have mined soil K reserves. Since most California soils were initially well supplied with potassium, K fertilization rates have historically been low. In areas of the state where irrigated production has been practiced for many decades, the cumulative removal of soil K amounts to thousands of pounds per acre in some fields. This has substantially reduced soil K availability and will force changes in fertilization practices in the future.

Soils generally contain huge quantities of K (fig. 6.1). The vast majority of it is locked in the primary minerals that make up the parent material; this "structural" K is unavailable for plant uptake. However, weathering makes a portion of structural K plant available over a long term and plays an important role in the maintenance of soil K fertility. Depending on the soil parent material and the degree of weathering, a substantial quantity of K may be fixed (trapped between layers of 2:1 silicate minerals like vermiculite or smectite; Pettygrove et al. 2011). Fixed K is partially plant available and may contribute substantially to crop nutrition in some soils. Exchangeable K and K in soil solution are plant available.

Similar to the situation with soil P, an equilibrium exists between soluble, exchangeable, and fixed K. Potassium concentration in soil solution is limited, typically 10 to 30 ppm in California soils. As plants remove K from soil solution, it is replenished from the other K pools. Conversely, fertilizer K application initially increases soil solution K, which subsequently results in most of the applied K moving to the cation exchange or into fixation sites.

Because potassium is not very mobile in neutral to alkaline soils, K leaching is usually not a significant agronomic concern except in very sandy soil with low CEC. In strongly acid soil, K leaching can be substantial because aluminum ions (Al^{3+}) displace K from cation exchange sites. Thankfully, such soils are very rare in California.

Soil potassium availability is usually estimated by the standard ammonium acetate exchangeable cations test. The name of this test is somewhat misleading: in soils with a substantial amount of fixed K, up to 20 to 30 percent of K removed by ammonium acetate extraction may be fixed K. California soils typically range from 50 to 500 ppm exchangeable K. As a general rule, soil with less than 100 ppm exchangeable K requires K fertilization to guarantee peak production of any vegetable

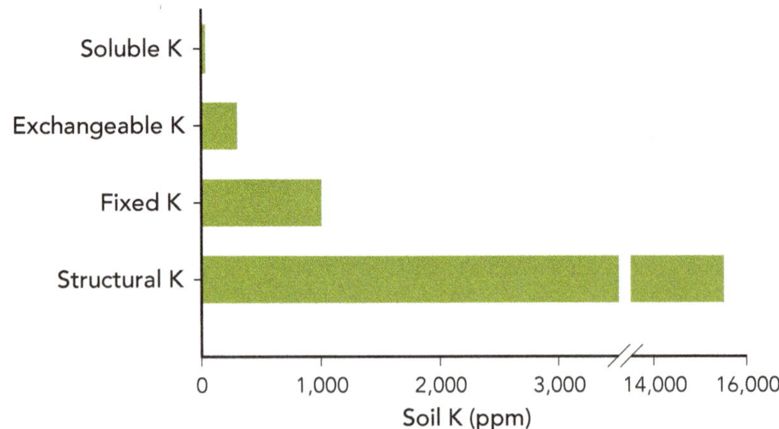

FIGURE 6.1.

Relative amounts of structural, fixed, exchangeable, and soluble K in a typical soil.

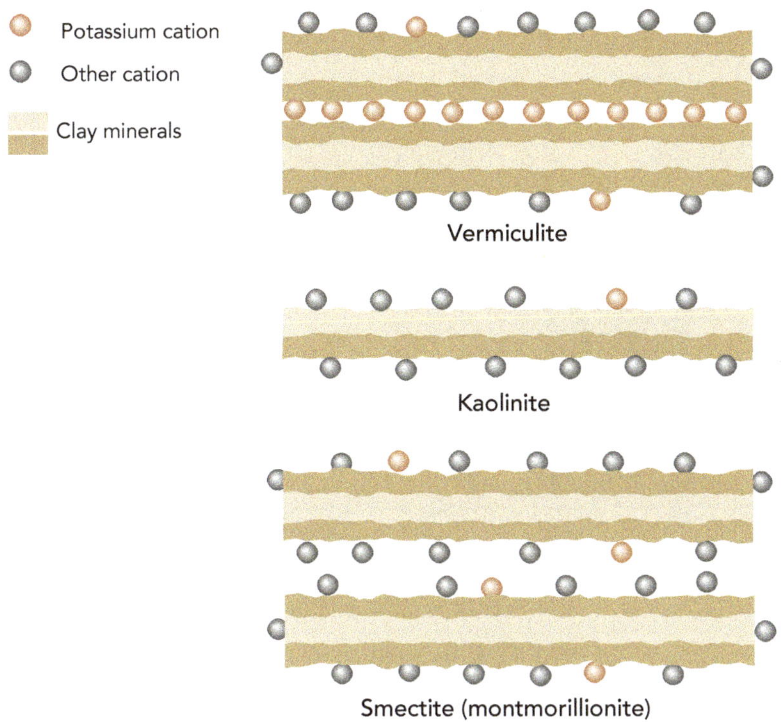

FIGURE 6.2.

Potassium fixation in different soil minerals.
Source: Pettygrove et al. 2011.

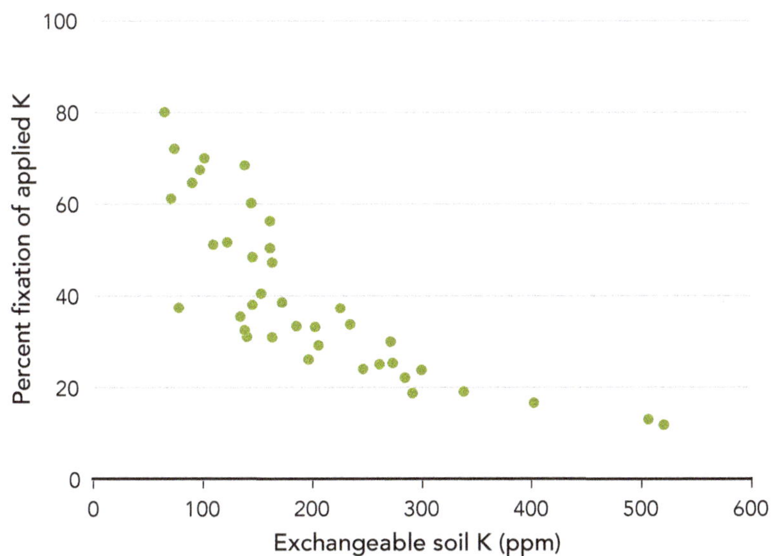

FIGURE 6.3.

Relationship between exchangeable K and K fixation potential in
Central Valley soils.
Source: Hartz et al. 2002.

crop, while some high-K-demand crops may respond to K fertilization in soils with more than 200 ppm (see table 3.6).

FACTORS AFFECTING SOIL POTASSIUM AVAILABILITY

Potassium Fixation

The ability of a soil to fix K depends on the prevalence of 2:1 silicate minerals like vermiculite or smectite in the soil, the soil texture, and the overall soil K status. Vermiculite and smectite are comprised of alternating sheets of different chemical structures (fig. 6.2). These sheets are loosely held together; K+ ions can get trapped, or fixed, between these layers. Potassium ions fixed in vermiculite are relatively firmed held, while those fixed in smectite can more easily move back into soil solution.

The presence and type of 2:1 minerals are a function of a soil's parent material and its degree of weathering. Potassium fixation is associated with soils formed from granitic parent material. This means that K fixation is mostly a phenomenon of Central Valley soils containing alluvium from the Sierra; coastal and Southern California soils are less likely to have a significant K fixation capacity. Determining the K fixation capacity of soil is not easy for a grower because commercial labs do not offer a dependable test. However, research on Central Valley soils show that K fixation is widespread (fig. 6.3). The forty soils in this figure were collected from representative tomato fields from Colusa to Fresno Counties. To estimate K fixation capacity, these soils were fertilized with a K solution, air-dried, then extracted to determine exchangeable K; the fraction of soluble K applied that could not be extracted after air-drying was considered to have been fixed. In general, the lower the exchangeable K, the greater the fixation capacity. Within a particular range of exchangeable K, the variability in fixation potential among soils represents the influence of the parent material (granitic material has higher fixation potential) and texture (silt-sized particles tend to have the highest fixation potential).

Efficient K fertilization can be challenging in soils with a high fixation capacity. The total amount of K that can be drawn into fixation sites can exceed several thousand pounds per acre, with more than 60 percent of applied K

FIGURE 6.4.

Uneven ripening of processing tomato fruit associated with low soil K availability or high soil Mg:K ratio. Photo: Timothy Hartz.

being fixed in the worst cases. It may not be economically feasible to completely overcome K fixation in such fields, but adjusting the application technique to minimize fixation can maximize the crop response to applied K. Cycles of soil wetting and drying increase K fixation, so preplant application presents the maximum opportunity for fixation. Banding may be more effective than broadcasting, as it limits the number of fixation sites the fertilizer K is exposed to. Potassium fertigation through drip irrigation delivers K to the concentrated root zone; frequent irrigation to minimize soil drying between irrigations may also reduce the amount of fertigated K that is fixed.

Cation Competition

Plants have mechanisms that allow the preferential uptake of certain nutrients from soil solution. This is necessary in order for plants to obtain enough K, which typically represents more than a third of all base cations (Ca, Mg, K, and Na) in plant tissue but usually constitutes less than 5 percent of base exchange in soil. These preferential uptake mechanisms are not perfect, and large imbalances between K and the other base cations in soil can limit K uptake. This is why not only the amount of soil exchangeable K but also the percentage of

base exchange that K represents is important. In general, when exchangeable K is more than 3 percent of base exchange (on a milliequivalent basis), cation competition is of little consequence. However, cation competition may increasingly be an issue as exchangeable K declines below 2 percent of base exchange.

An imbalance between exchangeable soil Mg and K appears to be particularly problematic. For example, a common disorder in processing tomato fruit is uneven ripening, in which the tissue at the stem end remains yellow after the rest of the fruit has turned red (fig. 6.4). This disorder, which is associated with low tissue K concentration, occurs most frequently in fields with a soil exchangeable Mg:K ratio greater than 10 on a milliequivalent basis (Hartz et al. 1999).

Rooting Density

Most plant K uptake comes from the soil solution; however, the soil solution holds only a small amount of K relative to crop K uptake. Consider the case of a celery crop, which in 1 day may take up more than 8 pounds of K per acre while transpiring 50,000 pounds (about 6,000 gallons) of water. If the soil solution K concentration is 20 ppm, the transpired water contains only 1 pound of K, or about 12 percent of the daily uptake. This demonstrates that mass flow is a minor influence on K uptake and that the vast majority of K taken up by the crop must move toward roots through diffusion (movement from areas of higher concentration toward lower concentration). Diffusive movement of K is quite slow in soil, less than 1 millimeter per day (Havlin et al. 2014). The greater the density of roots, the shorter the distance soil K must diffuse to come in contact with a root. Therefore, any factor (inadequate moisture, soil compaction, root disease, etc.) that limits root density also limits soil K availability.

Irrigation Method

Irrigation method can affect both the fraction of the soil volume that is kept moist enough for root activity and the soil depth at which K uptake occurs. Sprinkler irrigation wets the entire soil profile, in theory providing the maximum rooting volume. Furrow irrigation results in less-uniform wetting; for some

TABLE 6.1.

Common K fertilizers

Fertilizer	Chemical formula	% K_2O
potassium chloride	KCl	60–62
potassium nitrate	KNO_3	44–46
potassium sulfate	K_2SO_4	50–53
potassium thiosulfate	$K_2S_2O_3$	25

crops (such as melons) furrow irrigation is deliberately managed to keep the bed tops dry to maintain fruit quality. This may limit K availability, since K is typically most abundant in the top few inches of soil. Buried drip irrigation limits the volume of soil that roots can exploit and keeps the surface soil too dry for root activity through much of the growing season. Coupled with the potential for higher yields, buried drip irrigation can strain soil K resources. Figure 6.5 illustrates this fact: at the same level of exchangeable soil K, drip-irrigated tomatoes tend to have much lower fruit K concentration. This strongly suggests that the soil test K response threshold for buried drip irrigation is higher than it would be for other irrigation methods, and K fertilization requirements are therefore likely to be greater.

POTASSIUM FERTILIZERS

There are only a few commonly used K fertilizers (table 6.1). Once applied to soil, the K^+ ion is indistinguishable from one fertilizer source to another. Therefore, the choice of K fertilizer is driven primarily by cost, solubility, and the desirability of the companion anion. While all K fertilizers are considered water soluble, the amount of K that can be held in a concentrated solution varies among fertilizers, making some more practical for fertigation than others. Potassium thiosulfate is more soluble than KCl or KNO_3, which in turn are more soluble than K_2SO_4.

Companion anions can be important as nutrient sources (NO_3^- or SO_4^{2-}), for their effects on soil pH (thiosulfate is an acidifying agent), and for their potentially detrimental effects (Cl^-). Some growers are reluctant to use KCl out of fear that the chloride may harm their crops. While chloride can be toxic to vegetable crops at high concentration, some context is important. The amount of chloride in irrigation water is usually much greater than the amount added by fertilization with KCl. Potassium chloride application rates seldom exceed 200 pounds K_2O per acre per season, which would contain about 150 pounds of Cl per acre. Irrigation wells commonly have Cl concentration from 2 to 4 milliequivalents per liter, meaning that 1 acre-foot of irrigation water would contain from 200 to 400 pounds of Cl. Therefore, in fields where soil salinity is high and/or irrigation water Cl concentration is high, avoidance of KCl may be prudent; otherwise, use of KCl should present little risk.

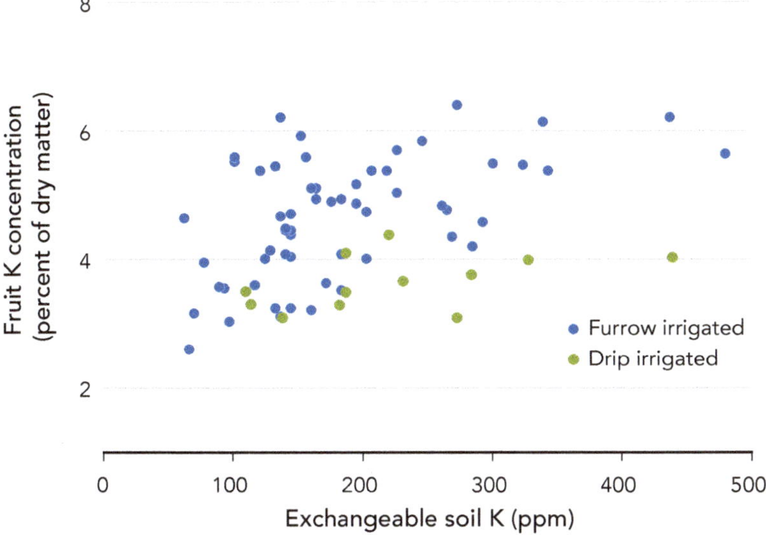

FIGURE 6.5.

Effect of irrigation method on K accumulation in processing tomato fruit.
Sources: Hartz and Bottoms 2009; Hartz et al. 1999, 2005.

POTASSIUM RATE AND TIMING

Soil test K level is the starting point for determining the appropriate K fertilization strategy for a given field, but other factors come into play. Approximate soil test K response thresholds are given in table 3.6. It must be recognized that these response thresholds are not infallible; the adequacy of soil K availability can also be affected by soil cation balance and factors that impact rooting density (irrigation pattern, soil structure, etc.). Potassium application is clearly warranted whenever the soil test level is lower than the response threshold and may be appropriate even when the soil test level is somewhat above the threshold. However, for soils far above the crop response threshold, K application is wasteful, in the sense that it increases luxury consumption (crop K uptake with no yield or quality benefit).

The importance of thoughtful K fertilizer management is illustrated by fig. 6.6. These data came from a study of coastal lettuce fields undertaken to determine leaf nutrient sufficiency guidelines; the preharvest leaf K sufficiency range established by this study

was 3 to 6 percent of dry weight. Participating growers tended to use a standard approach to K fertilization regardless of soil test K level. Over time, the effects of this approach are reflected in both the exchangeable soil K level and in the amount of K taken up by the crop. Growers for whom K application was routine tended to enrich the soil far above the soil test threshold for possible crop response (which is about 150 ppm K for lettuce). In such fields, crop leaf K concentration was far higher than needed to maximize yield and quality. Conversely, growers whose routine fertilization program did not include K application had, over time, reduced soil K availability; many of the unfertilized soils were below the possible crop response threshold, and some were below the leaf K sufficiency concentration.

Fields with soil test K substantially above the crop response threshold require no K application and may not require K for many years. Applying K to such fields is a waste of money. However, over the long term, growers must recognize that harvest K removal diminishes the soil fertility reserve, which is a tangible asset. As soil test K approaches the response threshold, K fertilization will be required to guarantee crop sufficiency. As a general rule, replacing the expected harvest K removal is appropriate; see table 2.1 for K removal estimates. Harvest removal may exceed 250 pounds of elemental K per acre for crops with high K demand (e.g., tomato, celery), which is far above typical application rates. While it is true that a lower K rate is often adequate to maximize the yield of the current crop, a rate substantially below the harvest K removal represents a future reduction in soil K availability.

Appropriate timing and method of K fertilization depend on field-specific conditions. Potassium leaching is a minor consideration; only in soil that is strongly acidic or has very low CEC (below 5 milliequivalents per 100 g) is there a substantial K leaching potential. This means that preplant application is often the technique of choice. Where the potential for soil K fixation is high, an application method that limits soil contact (e.g., banding) or limits wetting and drying cycles (e.g., drip fertigation) may result in more efficient uptake than

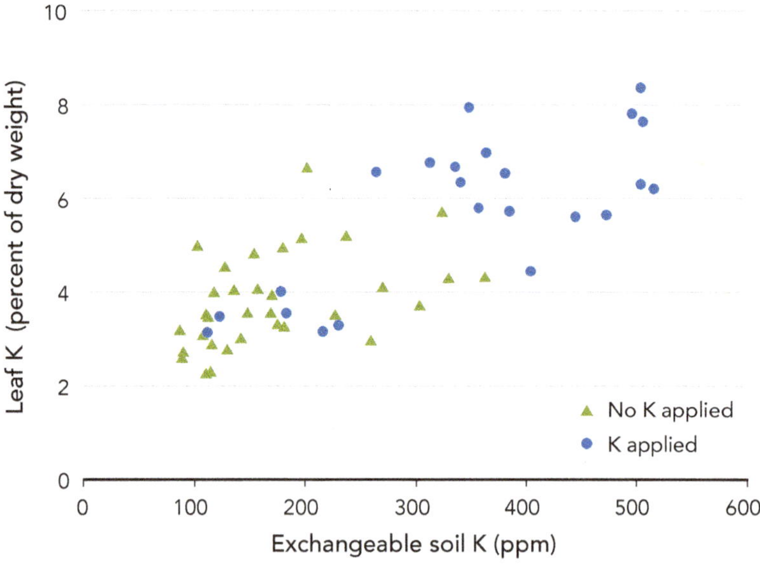

FIGURE 6.6.

Relationship between soil test K and preharvest leaf K concentration in lettuce; symbols identify whether fields received K fertilization.
Source: Adapted from Hartz et al. 2007.

a preplant broadcast application. Potassium application in the first half of the cropping season is likely to be more effective than later in the season. This is particularly true for fruiting crops (tomato, pepper, melon): K application before the end of fruit set maximizes yield potential, while delaying application until the fruit bulking period risks a reduction in fruit set.

Foliar K fertilization is a common practice for some growers, but the effectiveness of this practice is questionable. The amount of K that can be applied without foliar burning is limited (typically less than 5 pounds of K per acre per application, not all of which is absorbed into the leaf); by comparison, daily crop K uptake can exceed 6 to 8 pounds per acre. This means that repeat application is required to provide a meaningful amount of K. Across three field trials, Hartz et al. (1999, 2005) found that as many as six foliar K applications had no effect on tomato yield or quality.

POTASSIUM EFFECTS ON PRODUCE QUALITY

It is widely believed that K fertilization can improve produce quality, in particular the quality of fruit. While it is true that fruit K concentration may be correlated with quality factors such as color and soluble solids concentration, the practical ability to manipulate fruit quality with K fertilization is very limited. Many K fertilization trials on processing tomatoes have borne this out. Across more than twenty replicated field trials, K fertilization had no significant effect on fruit soluble solids concentration, even in fields that showed a fruit yield increase with K fertilization (Hartz et al. 1999, 2001, 2005). While a high rate of K fertilization (above 200 pounds of K per acre) did improve fruit color in some of these trials, the economic value of the improvement generally did not justify the fertilizer cost.

REFERENCES

Hartz, T. K., and T. G. Bottoms. 2009. Nitrogen requirement of drip-irrigated processing tomatoes. HortScience 44:1988–1993.

Hartz, T. K., P. R. Johnstone, D. M. Francis, and E. M. Miyao. 2005. Processing tomato yield and fruit quality improved with potassium fertigation. HortScience 40:1862–1867.

Hartz, T. K., P. R. Johnstone, E. Williams, and R. F. Smith. 2007. Establishing lettuce leaf nutrient optimum ranges through DRIS analysis. HortScience 42:143–146.

Hartz, T. K., E. M. Miyao, R. J. Mullen, and M. D. Cahn. 2001. Potassium fertilization effects on processing tomato yield and fruit quality. Acta Horticulturae 542:127–133.

Hartz, T. K., E. M. Miyao, R. J. Mullen, M. D. Cahn, J. G. Valencia, and K. L. Brittan. 1999. Potassium requirements for maximum yield and fruit quality of processing tomato. Journal of the American Society for Horticultural Science 124:199–204.

Hartz, T. K., C. Giannini, R. O. Miller, and E. M. Miyao. 2002. Estimating soil K availability for processing tomato production. Communications in Soil Science and Plant Analysis 33:1389–1400.

Havlin, J. L., S. L. Tisdale, W. L. Nelson, and J. D. Beaton. 2014. Soil fertility and fertilizers. Upper Saddle River, NJ: Pearson.

Pettygrove, S., T. O'Geen, and R. Southard. 2011. Potassium fixation and its significance for California crop production. Better Crops 95(4): 16–17.

Secondary Macronutrients: Calcium, Magnesium, and Sulfur

CALCIUM AND MAGNESIUM

Nearly all California soils contain sufficient plant-available Ca and Mg to supply vegetable crop nutrient requirements. In fact, these elements typically dominate the soil cation exchange to such an extent that crop Ca and Mg uptake is often far in excess of the amount required for normal plant growth and function. How can that be, when Ca-related disorders such as blossom end rot and tipburn are common? The answer lies in the complicated interplay of soil cation availability, plant physiology, irrigation management, and environmental conditions.

Soil Ca and Mg dynamics are similar to those of K, with the exception that there is no "fixation" of Ca or Mg as there can be with K. There is an equilibrium between Ca and Mg on cation exchange sites, in soil solution, and, in the case of Ca, in minerals of low solubility. As plants remove these nutrients from soil solution, the soil solution is replenished from these other pools. When Ca or Mg is added to the soil in the form of amendments or fertilizers, the soil solution may be temporarily enriched, but over time the equilibrium will shift back the other way.

Plant-available Ca and Mg are estimated using the exchangeable cations test. This test gives a relatively accurate picture of Mg availability. In the vast majority of California soils, Mg makes up at least 10 percent of estimated CEC, which should be more than adequate to supply plant needs. The picture for Ca is more complicated. In alkaline soils containing free lime, the exchangeable cations test can overestimate Ca availability by up to 20 to 30 percent by dissolving some calcium-containing minerals that would not be soluble in the field. However, the amount of exchangeable Ca in most California soils is so much larger than crop Ca uptake that it is clear that Ca fertilization should seldom be needed. To illustrate this point, consider that vegetable crops typically take up from 20 to 200 pounds of Ca per acre, depending on the crop and the soil characteristics. By comparison, it is unusual for soil to contain less than 4 milliequivalents per 100 grams of exchangeable Ca, which represents about 3,000 pounds of Ca per acre in the top foot of soil. Soil Ca could theoretically be limiting to crop growth

TABLE 7.1.

Comparative status of soil Ca and Mg (meq basis) by production region; values are means of at least 30 field soils per region

Region	Saturated paste Ca:Mg ratio	Exchangeable cations Ca:Mg ratio
Sacramento Valley	1.0	1.3
San Joaquin Valley	2.2	3.5
Salinas Valley	2.4	3.4

Sources: Data for Sacramento Valley and San Joaquin Valley from Hartz et al. 1999; data for Salinas Valley from Bottoms et al. 2013a, 2013b.

FIGURE 7.1.

Blossom end rot on chili peppers. Photo: Timothy Hartz.

in very sandy soil or in strongly acid soil in which the cation exchange is dominated by H^+ and Al^{3+}; such soils are exceedingly rare in California.

Regional differences exist in the relative amounts of soil Ca and Mg (table 7.1). San Joaquin Valley and Salinas Valley soils are dominated by Ca. Sacramento Valley soil has much higher levels of Mg (a legacy of serpentine parent materials) and consequently much lower Ca:Mg ratios. This regional variation in soil cation balance has soil fertility implications. As discussed in the previous chapter, cation competition between K and Mg affects the color uniformity of tomato fruit. On average, the Sacramento Valley has higher Mg and lower exchangeable K than the San Joaquin Valley and consequently has more tomato color disorders. The lower Ca:Mg ratio of Sacramento Valley soils also affects the relative plant uptake of those elements. While this has minimal effect on crop yield (both Ca and Mg availability are more than adequate to maximize growth), it can influence quality factors such as the firmness of fruit. Among the reasons that cantaloupes are not commonly grown in the Sacramento Valley is that experience has shown that fruit grown there are typically not as firm as those grown in San Joaquin Valley soils.

Calcium-Related Physiological Disorders

Two major categories of physiological disorders have a Ca connection: those affecting fruit and those affecting vegetative tissue. Blossom end rot (BER) of fruit (fig. 7.1) can affect tomatoes, peppers, and several types of cucurbits. BER occurs relatively early in fruit development when the rapidly expanding cells at the blossom end of developing fruit have insufficient Ca to form strong cell walls; subsequent cell death induces the visual symptom.

However, in California BER usually has little to do with soil Ca availability; rather, it is nearly always the result of water stress. Calcium is completely immobile in phloem (the vascular tissue through which nutrients are translocated from one plant part to another). This means that Ca can reach fruit only by transport in xylem (the vascular tissue that conducts water and soil nutrients from roots to aerial plant parts via transpiration). Fruit have lower transpiration rates than leaves and therefore lower Ca concentration; a tomato plant may have a leaf Ca concentration of 5 percent of dry weight, while the Ca concentration of fruit may be less than 0.15 percent of dry weight. Even a transient water stress that limits xylem flow into developing fruit can disrupt this tenuous Ca supply. Cells that are actively expanding when water stress occurs are easily damaged. One might expect BER to be more common in the Sacramento Valley than in the San Joaquin Valley based on regional differences in soil Ca. It is true that on average Sacramento Valley fruit has lower Ca concentration (Hartz et al. 1999), but water management is by far the more dominant influence on BER incidence.

BER can also be induced by heavy use of ammonium or urea fertilizers during fruit development: NH_4^+ ions can compete with Ca^{2+} for uptake and transport in xylem. However, this is seldom a significant factor in California fields, and normal use patterns of fertilizers such as UN-32 present minimal risk.

Calcium disorders that affect vegetative tissue include tipburn in lettuce (fig. 7.2) and black heart in celery. The mechanism

FIGURE 7.2.

Tipburn on romaine lettuce; damage typically occurs on inner leaves that have low transpiration rates. Photo: Richard Smith.

responsible for these disorders is somewhat different from BER. Leaves exposed to the sun and wind have high transpiration rates and therefore are generally well supplied with Ca. However, inner leaves shielded from sun and wind may transpire very little and consequently have difficulty obtaining sufficient Ca for normal growth. Water stress can play a significant role in the development of these disorders (particularly for black heart), but significant damage can occur even when adequate soil moisture is continuously maintained. Tipburn can be induced by field conditions in which temperature conducive to rapid growth is combined with low plant transpiration. For example, summer lettuce produced near the coast may exhibit more tipburn than lettuce grown farther inland. Although the inland sites are slightly warmer, coastal temperatures are still adequate for rapid growth, and coastal fog can significantly limit plant transpiration. In California, soil Ca status has been shown to have minimal effect on the occurrence or severity of lettuce tipburn (Hartz et al. 2007).

Calcium Fertilization

It is a common practice in vegetable production to apply soluble Ca fertilizers, usually through fertigation, typically at modest rates (less than 20 pounds Ca per acre per application). While fertilizers containing calcium nitrate may be used for the N content, the Ca content is widely thought to have fertilizer value; calcium thiosulfate or calcium chloride is clearly used for the Ca content. However, the efficacy of soluble Ca fertilizers in improving crop growth or reducing calcium-related physiological disorders under normal field conditions is questionable.

Given the high Ca content of typical California soils, fertigation of Ca at low rates could have a significant impact on the crop only if the fertigated Ca were more plant available than the soil Ca. It is true that all fertigated Ca is in the soil solution at least initially, while only a minority of exchangeable Ca is in the soil solution. However, soil solution Ca concentration is typically much higher than is commonly assumed. Hartz et al. (2007) collected soil from twenty vegetable fields in the Salinas Valley and the Central Valley and extracted the soil solution from them by high-speed centrifugation. The soil solution Ca concentration averaged more than 600 ppm, and all but one of these soils had more than 200 ppm. For context, hydroponic solutions used in greenhouse vegetable production, which are formulated to maximize crop growth and product quality, usually contain only 100 to 200 ppm Ca. It is also difficult to measure any plant response from low-rate Ca fertigation. Calcium fertigation, regardless of Ca source (calcium nitrate, calcium thiosulfate, or calcium chloride), did not significantly increase plant Ca uptake, reduce lettuce tipburn, or improve melon firmness (Hartz et al. 2007; Johnstone et al. 2008). Even at application rates of hundreds of pounds per acre, Misaghi et al. (1981) reported that soil-applied Ca fertilizers generally had no effect on lettuce tissue Ca content or on the severity of tipburn.

Foliar Ca application is similarly ineffective in minimizing calcium-related disorders, at least at normal application volumes. Since Ca is immobile in phloem, it cannot be translocated from one plant part to another. To be effective, a foliar Ca spray would have to not only reach the target plant part but also be effectively taken into the tissue. Getting good coverage of the inner leaves of lettuce or celery is unlikely unless very high spray volume is used; fertigating Ca through sprinklers may provide sufficient coverage to be useful. Obtaining good coverage on the blossom ends of fruit can also be challenging, and their waxy or netted surfaces present a barrier to effective absorption of applied Ca.

Calcium as a Soil Amendment

While low-rate application of Ca-containing fertilizers has minimal impact on the Ca nutrition of crops, higher-rate application (on the order of tons per acre) of Ca amendments can fundamentally alter both soil cation availability and soil physical properties such as soil crusting and water infiltration rate. Despite the fact that both Ca and Mg are divalent (double-charged) cations, Ca more effectively flocculates soil particles; Ca-dominated soils have better soil structure, are more resistant to soil crusting, and have better water infiltration than those dominated by Mg (Oster et al. 2016). However, even in Ca-dominated soil, infiltration problems can be induced by the use of irrigation water with very low salinity (< 0.7 dS/m; Ayers and Westcot 1985); such water can leach Ca from the top few centimeters of soil, destabilizing soil aggregates.

Soil crusting and infiltration are most often addressed by methods directed at modifying the top few inches of soil rather than the entire profile. In soil containing free lime, the application of acid materials can dissolve calcium carbonate, increasing soluble Ca at the soil surface. This is the principle behind the use of acid-based anticrustants to improve seedling germination. Gypsum (calcium sulfate, $CaSO_4 \cdot 2H_2O$) applied to the soil surface dissolves over several irrigation cycles or rainfall events, maintaining adequate Ca at the soil surface to improve infiltration. Soluble Ca from any source applied in irrigation water at as little as 3 milliequivalents per liter (60 ppm Ca) can measurably improve infiltration. However, this approach has limited residual effect and must be repeated for season-long improvement (Wildman et al. 1988, 1989).

SULFUR

Sulfur is usually abundant in California soils. The complex sulfur cycle in soil has an interplay of many factors, including atmospheric deposition, mineralization from soil organic matter, sulfate (SO_4^{2-}) leaching, and application of sulfur as a fungicide or in fertilizers, amendments, and irrigation water. However, in California's semiarid climate and soils with low organic matter, soil S availability is governed mostly by the amount of S applied with irrigation water, fertilizers, and amendments compared with the amount of sulfate lost to leaching. The S uptake requirement of vegetable crops ranges from about 20 to 80 pounds per acre, although in fields of high S availability actual uptake rates can be substantially higher. In the vast majority of California fields, that amount is dwarfed by the level of soil S availability.

Plants take up S from soil solution as SO_4^{2-}. If the soil solution contains only 1 milliequivalent per liter (16 ppm) SO_4-S, transpiration would usually bring enough S to the root surface to meet crop needs. Soil solution SO_4-S concentration exceeds 1 milliequivalent per liter in most California soils most of the time; therefore, the need to apply S to correct a deficiency is rare.

An important step in evaluating S sufficiency is to determine the SO_4-S concentration of the irrigation water: if the water contains more than 1 milliequivalent per liter, one can generally assume S to be sufficient in any field irrigated with it. Statewide, many irrigation wells exceed 1 milliequivalent per liter SO_4-S, but surface water sources generally have reasonably low sulfate concentration. A notable exception is the Colorado River, which is rich in sulfate.

Fields irrigated with low-sulfate water, particularly those low in soil organic matter (and therefore low S mineralization potential), may require S fertilization. That input can come from gypsum or a variety of fertilizers containing sulfates, thiosulfate, or sulfuric acid. When any of these products are used at normal rates, the amount of S applied is substantial in relation to crop S requirements. Elemental S is a "slow release" S source, in that it requires microbial activity to be oxidized to SO_4-S. The speed of that conversion depends mostly on particle size; it may take weeks to months for a substantial amount of SO_4-S to be generated from an elemental S application.

Soil testing for S availability is problematic, in that the time of sampling can have a large influence on the results. The common analytical procedures are an extraction test and saturated paste analysis, both measuring SO_4-S. The sulfate ion is highly mobile in soil and is susceptible to leaching. Soil samples taken in the fall, before winter rains, nearly always show substantial sulfate levels, but samples taken in spring, after winter rains, may show low S availability. In responding to a low soil S test, one must consider the context. Is the low value simply the result of heavy leaching? In the coming crop season, will a substantial amount of S be applied with the irrigation water or with macronutrient fertilizers or soil conditioners? Sulfur fertilization is required in relatively few cases.

Sulfur as a Soil Amendment

In addition to its role as a nutrient, S may also be used to acidify soil. Sulfuric acid (H_2SO_4) application to soil containing free lime creates a rapid neutralization reaction that releases soluble Ca; surface application of sulfuric acid may be used to reduce soil crusting. Elemental S can also be used to lower soil pH: microbial action in the soil oxidizes the S, releasing acidity in the process. Because microbial action is required, this reaction can take a considerable amount of time (weeks to months), depending on the particle size of the elemental S used: fine particles have greater surface area and react more quickly.

REFERENCES

Ayers, R. S., and D. W. Westcot. 1985. Water quality in agriculture. FAO Irrigation and Drainage Paper 29. Rome: U.N. Food and Agriculture Organization.

Bottoms, T. G., M. P. Bolda, M. L. Gaskell, and T. K. Hartz. 2013a. Determination of strawberry nutrient optimum ranges through DRIS analysis. HortTechnology 23:312–318.

Bottoms, T. G., T. K. Hartz, M. D. Cahn, and B. F. Farrara. 2013b. Crop and soil nitrogen dynamics in annual strawberry production in California. HortScience 48:1034–1039.

Hartz, T. K., P. R. Johnstone, R. F. Smith, and M. D. Cahn. 2007. Soil calcium status unrelated to tipburn of romaine lettuce. HortScience 42:1681–1684.

Hartz, T. K., E. M. Miyao, R. J. Mullen, M. D. Cahn, J. G. Valencia, and K. L. Brittan. 1999. Potassium requirements for maximum yield and fruit quality of processing tomato. Journal of the American Society for Horticultural Science 124:199–204.

Johnstone, P. R., T. K. Hartz, and D. M. May. 2008. Calcium fertigation ineffective at increasing fruit yield and quality of muskmelon and honeydew in California. HortTechnology 18:685–689.

Misaghi, I. J., C. A. Matyac, and R. G. Grogan. 1981. Soil and foliar applications of calcium chloride and calcium nitrate to control tipburn of head lettuce. Plant Disease 65:821–822.

Oster, J. D., G. Sposito, and C. J. Smith. 2016. Accounting for potassium and magnesium in irrigation water quality assessment. California Agriculture 70:71–76.

Wildman, W. E., W. H. Krueger, and R. E. Pelton. 1989. Calcium amendments for water penetration in flooding systems. California Agriculture 43(3): 14–15.

Wildman, W. E., W. L. Peacock, A. M. Wildman, G. G. Goble, J. E. Pehrson, and N. V. O'Connell. 1988. Soluble calcium compounds may aid low-volume water application. California Agriculture 42(6): 7–9.

8
Micronutrients

Vegetable crops take up very small quantities of micronutrients, usually less than 1 pound per acre per micronutrient. Soils often contain reasonably large quantities of these elements, but only a small fraction is plant available. The cationic micronutrients Zn, Fe, Mn, and Cu are very sparsely represented on the soil cation exchange. Their plant availability is governed by an equilibrium between soil solution, precipitated and adsorbed forms, and chelation by soluble organic compounds. These soluble compounds can be root exudates or products of the breakdown of plant residue or soil humic matter. Soil solution concentration of cationic micronutrients, which is low even in soil of neutral pH, is reduced by high soil pH and by high bicarbonate irrigation water. Under these conditions the role of the soluble organic chelates is particularly important: chelation limits precipitation of these ions and increases their rate of diffusion toward roots. This is one reason the application of organic matter (manure, compost, cover crops, etc.) tends to improve micronutrient availability. However, in peat soils organic matter chelation of micronutrients can result in limited plant availability, particularly for Cu.

The factors governing B availability are quite different. Boron exists in soil solution as undissociated boric acid (H_3BO_3) or in several anionic forms; the soil solution concentration is usually high enough to supply crop needs. Boron can be leached from soil as either an uncharged compound or an anion. However, it can be complexed by soil organic matter or adsorbed onto the surface of certain soil minerals; this supply of B is in equilibrium with soil solution and can slowly buffer leaching losses. Fields that receive high winter rainfall or are irrigated with very low-B water can develop B deficiency. Where irrigation water contains at least 0.2 ppm B, leaching losses are constantly replaced and deficiency is unlikely.

Molybdenum is a rare element in soil; total Mo concentration is often less than 1 ppm, only a fraction of which is plant-available. The concentration of plant-available Mo is generally too low to be accurately measured by a practical soil test, which means that Mo deficiency is best assessed by tissue analysis. The MoO_4^{2-} ion is the primary form present in soil solution; its concentration declines sharply with decreasing soil pH, meaning that

Mo deficiency is most likely in acid soils. Molybdenum deficiency in California vegetable production is rare but can occur. Gubler et al. (1982) documented Mo deficiency in melons produced on acid soils (pH 5.3 to 6.3) in the Sacramento and San Joaquin Valleys; foliar Mo applications corrected the deficiency.

Deficiencies of Zn, Fe, Mn, and Cu are relatively rare in California vegetable production. This may seem surprising, given that high soil pH is known to reduce micronutrient availability. Table 8.1 compares micronutrient levels measured in soils collected from vegetable fields around the state; these data represent more than 100 fields from each production area. The generally higher values in the coastal areas may be a result of more intensive fertilization practiced over many years. The lower values in the San Joaquin Valley and the Imperial Valley may relate to generally higher pH and lower soil organic matter than in the other production areas; soil Fe and Mn availability in particular decline with increasing pH and decreasing organic matter.

The percentage of California fields rated as likely to respond to micronutrient fertilization depends on the critical soil value used. Since little research into micronutrient issues on vegetable crops has been done in California in recent decades, no definitive soil critical values have been developed here. Existing references suggest widely varying soil critical levels, and variation among different crops in sensitivity to soil micronutrient availability makes it difficult to establish a broadly applicable soil critical level. The soil critical levels listed in table 8.1 are a synthesis of several reference sources and represent the levels below which micronutrient availability is thought likely to limit growth, particularly for crops that are sensitive to deficiency (table 8.2). Sensitive crops may respond to micronutrient fertilization even in fields with soil test values higher than those listed here.

The inexact nature of soil micronutrient critical levels can be illustrated by the fact that high yields of lettuce, onions, and potatoes, crops purported to be sensitive to Mn deficiency, are routinely achieved in soils near or even below the critical level given in table 8.1. Conversely, a study undertaken to establish lettuce leaf nutrient sufficiency concentrations (Hartz et al. 2007) found that undesirably low leaf Cu concentration commonly occurred even in fields with extractable Cu levels up to 2 ppm.

The bottom line is that although documented cases of growth-limiting micronutrient deficiency are rare in California vegetable production, a substantial number of fields appear to have marginal availability of one or more micronutrients. Given the generally high value of vegetable crops and the modest cost

TABLE 8.1.

Regional differences in DTPA-extractable soil micronutrients and frequency of potential soil micronutrient deficiency

Production region	Median value (ppm)	Critical level* (ppm)	% of fields below critical level
Zn			
Imperial Valley	1.0	0.5	6
San Joaquin Valley	1.7		4
Sacramento Valley	1.5		6
Coastal regions	2.5		1
Fe			
Imperial Valley	13	5.0	2
San Joaquin Valley	9		8
Sacramento Valley	33		0
Coastal regions	17		0
Mn			
Imperial Valley	3	1.5	24
San Joaquin Valley	3		18
Sacramento Valley	7		5
Coastal regions	13		2
Cu			
Imperial Valley	1.3	0.5	6
San Joaquin Valley	1.2		3
Sacramento Valley	1.7		5
Coastal regions	1.8		0

Sources: Brown and deBoer 1983; Havlin et al. 2014; Horneck et al. 2011; Ludwick et al. 2010.

*Value below which crop response to fertilization is likely; sensitive crops may require somewhat greater soil availability.

TABLE 8.2.

Vegetable crops with high sensitivity to micronutrient deficiency

Zinc	Manganese	Iron	Copper	Boron	Molybdenum
beans	beans	beans	beets	beet	broccoli
garlic	lettuce	broccoli	carrots	broccoli	cauliflower
onion	onion	cauliflower	lettuce	cauliflower	lettuce
sweet corn	potato	spinach	onion	celery	onion
	spinach		spinach		spinach
	tomato				

Source: Adapted from Havlin et al. 2014.

TABLE 8.3.

Common micronutrient fertilizers

Element	Fertilizer	Approximate analysis (elemental %)	Typical application rate (elemental lb/ac)
zinc	zinc sulfate	36	2–10
	zinc oxide	78	5–15
	chelated zinc	6–14	0.2–2.0
iron	chelated iron	5–14	0.5–1.0
manganese	manganese sulfate	25	5–40
	chelated manganese	5–12	0.2–0.5
copper	copper sulfate	25	2–10
boron	borax	11	1–3
	Solubor	20	1–3
molybdenum	sodium molybdate	40	0.1–1.0
	ammonium molybdate	54	0.1–1.0

of micronutrient fertilizers, it may be prudent to fertilize soils with marginal soil test values to eliminate the possibility of deficiency. Combining soil analysis and leaf analysis (see chapter 9) may help identify fields of marginal micronutrient availability.

A variety of fertilizer materials are available when micronutrient application is warranted. The most common soil-applied sources of Zn, Mn, and Cu are their sulfate salts (table 8.3), which are quite soluble and provide immediate availability. Zinc oxide is also used; it has very low solubility but over time does provide substantial Zn availability. Soil-applied sources of Fe are generally not very effective because soluble Fe precipitates rapidly in high-pH soil. A range of chelated micronutrient fertilizers are available and can be highly effective. However, their cost per unit of micronutrient is much higher than the inorganic sulfate fertilizers, so the use of chelated micronutrients is typically reserved for foliar application. Foliar application is a more practical delivery technique for micronutrients than for macronutrients, because a single foliar spray can supply the equivalent of a crop's seasonal micronutrient uptake. Boron sources vary in solubility: granular borax is the least-soluble common fertilizer material, and Solubor is the most soluble. Soil-applied and foliar B fertilization can be effective.

Chlorine availability is never growth limit-

TABLE 8.4.

Relative tolerance of vegetable crops to boron in irrigation water

Irrigation water B concentration range (ppm)				
10–15	*4–6*	*2–4*	*1–2*	*0.5–1.0*
asparagus	beet	artichoke	broccoli	bean
	parsley	cabbage	carrot	garlic
	tomato	cantaloupe	cucumber	onion
		cauliflower	pea	
		celery	pepper	
		lettuce	potato	
		sweet corn	radish	

Source: Maas 1986.

ing in California vegetable production, but crop Cl toxicity can be an issue. Similarly, B toxicity is a more common concern than B deficiency. Where excessive amounts of these elements accumulate in soil, the source is usually irrigation water. Vegetable crops differ widely in their sensitivity to these elements. Most vegetables can tolerate irrigation water Cl concentration up to 10 milliequivalents per liter (350 ppm) with minimal problem, provided a reasonable leaching fraction is maintained to prevent soil buildup (Ayers and Westcot 1985). Relative crop B tolerance ratings are given in table 8.4. These ranges were developed in sand culture experiments and represent the maximum irrigation water B concentration tolerated without yield loss. In field soils, where significant B accumulation may occur, crop tolerance would likely be somewhat lower.

REFERENCES

Ayers, R. S., and D. W. Westcot. 1985. Water quality in agriculture. FAO Irrigation and Drainage Paper 29. Rome: U.N. Food and Agriculture Organization.

Brown, A. L., and G. J. deBoer. 1983. Soil tests for zinc, iron, manganese and copper. In H. M. Reisenauer, ed., Soil and plant tissue testing in California. Oakland: University of California Agriculture and Natural Resources Publication 1879.

Gubler, W. D., R. G. Grogan, and P. P. Osterli. 1982. Yellows of melons caused by molybdenum deficiency in acid soil. Plant Disease 66:449–451.

Hartz, T. K., P. R. Johnstone, E. Williams, and R. F. Smith. 2007. Establishing lettuce leaf nutrient optimum ranges through DRIS analysis. HortScience 42:143–146.

Havlin, J. L., S. L. Tisdale, W. L. Nelson, and J. D. Beaton. 2014. Soil fertility and fertilizers. Upper Saddle River, NJ: Pearson.

Horneck, D. A., D. M. Sullivan, J. S. Owen, and J. M. Hart. 2011. Soil test interpretation guide. Corvallis: Pacific Northwest Cooperative Extension Publication EC 1478.

Ludwick, A. E., L. C. Bonczkowski, M. H. Buttress, C. J. Hurst, S. E. Petrie, I. L. Phillips, J. J. Smith, and T. A. Tindall. 2010. Western fertilizer handbook. 9th ed. Long Grove, IL: Waveland Press.

Maas, E. V. 1986. Salt tolerance of plants. Applied Agricultural Research 1:12–26.

In-Season Nutrient Monitoring

TISSUE ANALYSIS

Measuring the concentration of essential elements in plant tissue is a well-established practice. Unlike the semi-quantitative "indexing" tests for soil nutrients, tissue analysis quantitatively measures the total amount of the nutrients or nutrient forms present. Two types of tissue analyses are commonly used to monitor vegetable crops: measurement of NO_3-N, phosphate-phosphorus (PO_4-P), and K concentration in petioles or midribs; and measurement of total nutrient concentrations in leaves. Leaf total nutrient analysis measures all chemical forms of a nutrient. For N and P that includes both the mineral forms measured in petiole analysis and N and P bound in organic compounds like chlorophyll or amino acids. Micronutrient concentrations are usually evaluated using leaf analysis.

Sample Collection and Handling

Diagnostic accuracy requires that tissue samples be representative of the field. This can be accomplished by collecting a minimum of twenty to thirty leaves from representative plants around the field; depending on field size and size of the leaf or petiole, fifty or more leaves may have to be collected to form a good sample. Leaf position on the plant is also important. Mobile elements like N, P, and K tend to decline in concentration with advancing leaf age because these nutrients can be exported from leaves to developing fruit or other storage organs. By contrast, immobile elements like Ca tend to accumulate in leaves over time due to continuing supply in the transpirational stream without export from leaves (Ca does not move through the phloem; see chapter 7).

The tissue usually recommended for sampling is recently mature leaves, which are found several nodes back from a growing point. In vegetative crops that form a head (e.g., iceberg lettuce), the leaf to collect is the "wrapper" leaf (the outermost leaf associated with the developing head). Choosing which leaf to sample can be a judgment call; luckily, leaves within 1 or 2 nodes on either side of recently mature or wrapper leaves have similar nutrient concentrations. When collecting leaf samples for total nutrient analysis, it usually makes relatively little difference whether the petioles or midribs are removed from the leaf blades. In most crops the petiole constitutes only a small fraction of the leaf dry weight, and petiole total nutrient concentrations tend to be relatively similar to that of the blade tissue. Of course, if the evaluation standards being used explicitly state "whole leaf" or "leaf blade," follow those guidelines.

Correct identification of the crop growth stage is also critical for accurate diagnosis because the N, P, and K concentration of recently mature leaves may decline substantially as the growing season progresses. This decline is particularly significant in fruiting crops (e.g., tomatoes, peppers, and melons), and tuber crops (e.g., potatoes). These nutrients are translocated from leaves to developing fruit or tubers; in the process, leaves can be drained of much of their content of these mobile elements. The fruiting habit of the plant can affect the steepness of the decline in leaf nutrient concentration. For example, determinate processing tomato varieties, by virtue of their concentrated fruit setting characteristic, show a steeper seasonal decline in leaf nutrient concentration than do indeterminate fresh market tomato varieties.

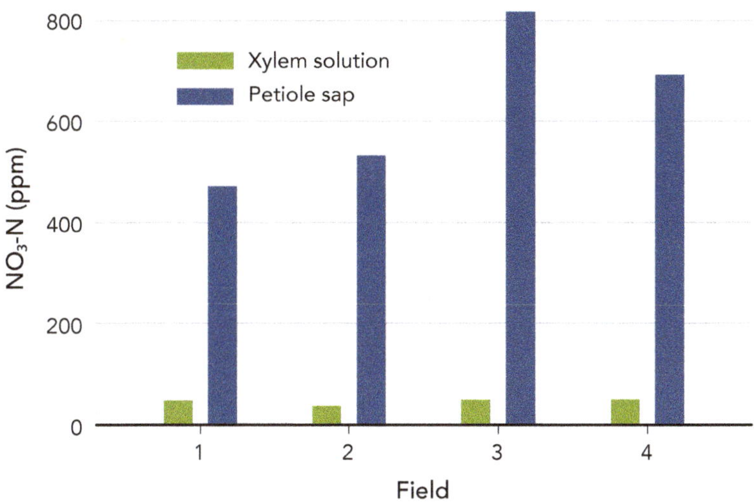

FIGURE 9.1.

Comparison of NO$_3$-N concentration in xylem solution and petiole sap of processing tomatoes during early fruit set.
Source: Hartz 2009.

Sample handling is important for petiole samples because NO$_3$-N and PO$_4$-P concentrations tend to decline with time after collection. This is because the normal cellular processes that convert these mineral nutrient forms into organic compounds continue to function until the petioles are either refrigerated or dried. Promptly oven-drying petiole samples after collection is ideal. Alternatively, keeping petiole samples cold until delivery to the lab helps limit postcollection changes in NO$_3$-N and PO$_4$-P concentrations. Handling leaf samples for analysis of total nutrient concentration is less exacting because the lab analysis measures all nutrient forms, whether mineral or organic. Air-drying stabilizes leaf samples if prompt lab delivery is not practical. If micronutrients are to be measured, remove dust from the leaves before drying. This can be done by washing leaves in water containing a small amount of phosphate-free detergent, then rinsing them in deionized water. Surface dust can increase the concentration of micronutrients such as iron.

Interpretation of Results

Interpretation of tissue nutrient concentration is fraught with potential errors. To minimize errors in interpretation, it is important to understand some basic plant physiology and to recognize the limitations of published diagnostic standards. Petiole analysis has historically been regarded by many in the vegetable industry as giving the best indication of plant macronutrient status. The very idea of analyzing the petiole (which contains the xylem vessels conveying soil nutrients to the leaves) instead of analyzing leaf samples suggests that the measurement is in some way related to the amount of soil nutrients currently being taken up. On close examination this assumption does not hold up.

Hartz (2009) used a pressure chamber to collect the solution inside xylem vessels of tomato petioles. The NO$_3$-N concentration of that solution was compared with the NO$_3$-N concentration of matching samples of sap obtained by crushing tomato petioles. Averaged across samples from four different fields collected at the early fruit set growth stage, sap NO$_3$-N was more than 10 times higher than in the xylem solution (fig. 9.1). This large difference suggested that the vast majority of NO$_3$-N in petiole tissue is actually contained in plant cells, with only a small fraction of it currently in the transpirational stream. This is an important distinction, because the rate at which NO$_3$-N and PO$_4$-P contained in plant cells are converted from mineral forms into complex organic compounds like chlorophyll or amino acids can be affected by environmental variables (e.g., temperature, light, and soil moisture) and by the demand for nutrients by developing fruit. The confounding influence of these factors weakens the relationship between petiole NO$_3$-N or PO$_4$-P concentration and the current soil availability of those nutrients. Since environmental conditions can change markedly from day to day, petiole nutrient concentration can fluctuate significantly. Petiole NO$_3$-N and PO$_4$-P concentrations can fluctuate by 20 percent or more from one day to the next, even while soil nutrient availability remains essentially unchanged.

A number of recent studies in California and elsewhere have suggested that petiole NO$_3$-N concentration is too unreliable a measurement upon which to base fertilization decisions. Hartz and Bottoms (2009) documented that widely used tomato petiole NO$_3$-N sufficiency standards are much higher than required for optimal production, particularly after fruit set. Sanchez (1998) conducted several dozen N fertilization trials in lettuce, broccoli, and cauliflower in the Imperial Valley/Yuma production area. He concluded that basing sidedress N application on petiole NO_3-N concentration was an unreliable technique that led to unnecessary fertilization about 30 percent of the time. Bottoms et al. (2013) reported that the pattern of petiole NO_3-N observed over the production season in high-yield strawberry fields was so variable as to make petiole monitoring worthless as a diagnostic tool for N management.

The bottom line is that petiole NO_3-N and PO_4-P testing has marginal value. Nutrient concentrations well above established standards may provide assurance that soil macronutrient availability is not currently growth limiting. However, nutrient concentrations below established standards cannot reliably be interpreted as showing low soil nutrient availability or the need for additional fertilization.

Petiole sap testing has been proposed as an on-farm alternative to lab analysis of petiole tissue. In this approach, petioles are crushed and the nitrate concentration of the sap is determined using a handheld ion-selective electrode. Unfortunately, petiole sap monitoring suffers from the same limitations as lab testing of dry petiole tissue, plus additional measurement variability related to calibration of the electrode, interference from other ions in the sap, and variation in tissue moisture content. Given the serious flaws of petiole analysis, whether of dry tissue or sap, no petiole nutrient sufficiency guidelines are given in this publication.

In contrast to petiole testing, leaf total nutrient concentration has consistently been shown to be a valid measure of the current plant nutrient status. Leaf total nutrient concentrations are also reasonably stable, seldom changing more than 10 to 15 percent within a week. Diagnostic standards for leaf macronutrient sufficiency are available from a variety of reference sources, but unfortunately there is considerable variability among these sources as to the recommended leaf nutrient levels. Part of this variability is semantic: some sources give critical levels (nutrient concentrations below which growth is likely to be impaired), while others list sufficiency ranges (nutrient concentration ranges typical of high-yield fields).

Table 9.1 lists macronutrient sufficiency ranges for major California vegetable crops. Leaf nutrient concentrations within these ranges are adequate to support optimal growth. A nutrient concentration below the sufficiency range does not always indicate deficiency, but the farther the nutrient concentration falls below the sufficiency range, the more likely that nutrient is or will soon become growth limiting. Leaf nutrient concentrations above these ranges may indicate excessive soil nutrient availability. It may also indicate that there is another factor limiting crop productivity: for example, if excessive heat prevents good fruit set in tomatoes, leaf nutrient concentrations may not decline in the usual manner, as there is little demand for nutrient translocation to developing fruit.

There is a degree of consistency in the leaf nutrient concentration ranges among botanically related crops. One can generally assume that leaf macronutrient sufficiency ranges for a crop not listed in table 9.1 are similar to those of a closely related crop in the table. There are exceptions to this rule: for example, the leaf N sufficiency concentration in cabbage is substantially lower than in broccoli or cauliflower (Smith et al. 2016).

There is limited California research upon which to set vegetable crop leaf micronutrient sufficiency ranges, and the sufficiency ranges given by sources vary widely. Part of this variability arises from the fact that micronutrient sufficiency ranges have often been based on the leaf concentrations commonly observed in a particular production region rather than on actual fertilizer trials where deficiency or sufficiency has been documented. Since production regions may have radically different soil characteristics, climates, and production systems, the typical leaf micronutrient concentrations observed may differ substantially between production regions. However, by evaluating published micronutrient sufficiency ranges across a number of vegetable crops we can infer a minimum leaf concentration below which growth may be adversely affected:

TABLE 9.1.

Macronutrient sufficiency ranges for recently mature whole leaves

Crop	Growth stage	N	P	K
		Leaf sufficiency range (% of dry weight)		
baby greens	preharvest	4.0–6.0	0.40–0.80	5.0–8.0
broccoli	buttoning	3.5–5.0	0.30–0.70	2.5–4.0
carrot	midgrowth	2.5–3.5	0.25–0.50	2.5–4.0
	preharvest	2.0–3.0	0.20–0.40	1.5–3.0
cauliflower	midgrowth	4.0–6.0	0.40–0.70	3.0–4.5
	preharvest	3.0–4.5	0.40–0.70	2.0–4.0
cantaloupe and honeydew	fruit set	3.0–4.5	0.30–0.60	2.5–4.0
	preharvest	2.5–4.0	0.25–0.50	2.0–4.0
lettuce (head and romaine)	early heading	4.0–5.5	0.40–0.70	3.5–6.0
	preharvest	3.5–4.5	0.35–0.60	3.0–6.0
onion	bulb initiation	2.5–4.0	0.25–0.40	2.5–4.0
pepper (bell and chili)	early fruit set	3.5–5.0	0.35–0.60	2.5–4.0
	first harvest	2.5–3.5	0.25–0.40	2.0–3.5
potato	before tuber initiation	4.0–6.0	0.35–0.60	3.5–6.0
	tuber bulking	2.5–3.5	0.20–0.40	2.0–5.0
tomato (determinate)	early flower	4.0–5.0	0.35–0.50	2.5–3.5
	full bloom	3.5–4.5	0.30–0.50	2.0–3.0
	early red fruit	2.7–3.5	0.25–0.40	1.2–2.5
watermelon	fruit set	3.0–4.0	0.30–0.60	3.0–5.0
	first harvest	2.5–3.5	0.25–0.40	2.0–4.0

Sources: Hartz 2006; Hartz et al. 1998, 2007; Hochmuth et al. 2012; Jones et al. 1991; Ludwick et al. 2010; Reuter and Robinson 1997.

- 5 ppm Cu
- 20 ppm B and Zn
- 25 ppm Mn
- 30 ppm Fe

Leaf micronutrient concentrations above these levels can generally be assumed to be adequate for crop growth. Leaf Mo concentration above 0.5 ppm is adequate for most vegetable crops, with 0.2 ppm adequate for lettuce, pepper, spinach, and tomato (Hochmuth et al. 2012).

Leaf micronutrient levels are often several times higher than these minimum concentrations, particularly for Fe. The only micronutrient toxicity that occurs with any regularity is B. Boron toxicity is indicated by leaf B concentration above about 100 ppm for sensitive crops or 200 ppm for tolerant crops (see table 8.4), combined with symptoms consistent with B accumulation, such as interveinal chlorosis and edge burning of older leaves.

Leaf sufficiency ranges for the secondary macronutrients (Ca, Mg, and S) are likewise difficult to determine, in part because deficiency of these elements is so rare. Leaf Mg and S concentration above 0.3 percent can generally be assumed to be adequate; in California vegetable production, leaf concentrations of these nutrients are often much higher. Leaf Ca concentration varies widely among crops and between fields of a given

TABLE 9.2.

Common symptoms of crop nutrient deficiency

Nutrient		Deficiency symptoms
N	nitrogen	General chlorosis, appearing on older foliage first, can progress to the whole plant; new growth may be spindly.
P	phosphorus	Initial symptom is slow growth, which may not be obvious; with severe deficiency, foliage may turn dull or dark green, or even purple, particularly on veins and undersides of leaves.
K	potassium	Margins of older leaves become tanned and scorched, with necrotic spots developing under severe deficiency.
Ca	calcium	Vegetative growing points or actively growing fruit or other storage organs (tubers, roots) become necrotic; affected tissue may be colonized by plant pathogens.
Mg	magnesium	Interveinal chlorosis on older leaves, progressing to marginal necrosis if severe; leaves may take on a reddish-purple color.
S	sulfur	Young leaves take on a uniform yellow cast, distinct from the interveinal chlorosis that characterizes Fe or Zn deficiency.
Zn	zinc	Interveinal chlorosis and marginal necrosis of young leaves; leaf size may be substantially reduced.
Fe	iron	Interveinal chlorosis of young leaves; does not typically progress to necrosis.
Mn	manganese	Interveinal mottling or chlorosis on young leaves progressing to necrosis; new growth may be stunted.
Cu	copper	Stunting and malformation of young leaves, chlorosis or bleaching possible.
B	boron	Death of terminal buds; young leaves distorted, necrotic; stems may be rough, cracked.
Mo	molybdenum	General yellowing or interveinal chlorosis of older leaves; marginal necrosis and bleaching of leaf blades may develop.

crop, based on the soil Ca status and the production environment (higher transpiration results in higher leaf Ca concentration). For crops in which recently mature leaves are fully exposed to solar radiation and therefore have high transpiration rates, leaf Ca concentration typically ranges from approximately 1 to 3 percent. However, leaf Ca concentration as low as 0.6 percent (cauliflower, Hartz 2006) or 0.5 percent (romaine lettuce, Hartz et al. 2007) appears to be adequate for good growth, as the sampled leaves in these crops have lower transpiration rates due to their protected position on the plant. Leaf Ca concentration can be highly variable and difficult to interpret; soil analysis is the more dependable technique for assessing Ca status.

DIAGNOSING NUTRIENT DEFICIENCY SYMPTOMS

Nutrient deficiencies often produce characteristic visual symptoms on affected plants (table 9.2). However, diagnosing nutrient deficiencies solely by visual symptoms is unreliable because other conditions unrelated to nutrient status can confound diagnosis. For example, leaf chlorosis, scorching, and necrosis, symptoms often associated with nutrient deficiency, can also be caused by herbicide damage, plant pathogens, water stress, salinity buildup, or even nutrient toxicity; soil and/or tissue testing is required to confirm a nutrient deficiency. Foliar symptoms can be useful in focusing attention on particular nutrients.

When assessing visual symptoms, pay particular attention to the location on the plant where symptoms begin. Because N and K are highly mobile in the plant, older leaves can translocate these nutrients to new growth, causing symptoms to develop first on the older leaves. Conversely, most micronutrients are less mobile in plants, and deficiency symptoms typically show first, and most clearly, on new growth.

It should be emphasized that growth-limiting nutrient deficiency is uncommon in California vegetable production, given the generally high soil fertility maintained by most vegetable growers. Foliar symptoms consistent with nutrient deficiency are more often induced by factors unrelated to soil fertility, such as virus infection, salinity buildup, or pesticide phytotoxicity.

SOIL TESTING

Phosphorus, potassium, and most micronutrients have complex equilibrium reactions in soil that result in relatively stable soil availability over time; in-season soil analysis is likely to yield results very similar to preplant analysis. Therefore, leaf analysis is the primary in-season monitoring tool to document sufficiency of these elements. Nitrogen is unique in that its soil availability can fluctuate widely over a relatively short period, so either leaf analysis or in-season soil nitrate testing could be used to guide N fertilization. Leaf N concentration well below the sufficiency range does suggest that soil N supply is not keeping up with plant N demand and fertilization is justified. However, leaf N concentration within the sufficiency

range provides little information about current soil N availability and is therefore of limited value in predicting future fertilization requirements. This is because unless plant N uptake is straining the soil N supply, the correlation between leaf N concentration and soil N availability tends to be poor.

Consider the case of N management in lettuce production. In a study of more than twenty commercial fields sampled at the early heading stage, there was no correlation between leaf N and soil NO_3-N across a wide range of soil NO_3-N concentrations (fig. 9.2). Leaf N concentration provided no information on current soil N availability and therefore provided no practical guidance on whether additional N application was necessary.

By contrast, in-season soil nitrate testing has been shown to be of great value in managing vegetable crop N fertilization. The importance of in-season soil nitrate testing is apparent when one contemplates the enormous range of residual soil NO_3-N observed in California vegetable fields. A recent study of processing tomato fields found that residual soil nitrate varied from less than 50 to more than 300 pounds of N per acre in the top 20 inches of soil (Lazcano et al. 2015). Similarly, Bottoms et al. (2012) and Smith et al. (2016) reported that residual soil NO_3-N varied among coastal vegetable fields by hundreds of pounds per acre. Clearly, using a standard N fertilization program in all fields, without consideration of the amount of residual soil nitrate present, is highly inefficient.

The timing of soil nitrate sampling is important. Soil samples taken before planting may not accurately reflect soil NO_3-N concentration after crop establishment, because nitrate may be leached by irrigation applied to establish the crop. Also, where a crop residue or compost has been incorporated shortly before planting, preplant sampling would miss the N mineralized in the weeks between incorporation and the first in-season N application. Therefore, it is desirable to sample as close to the first postplant N application as is possible. Soil sampling to inform in-season N management is often referred to as pre-sidedress soil nitrate testing (PSNT);

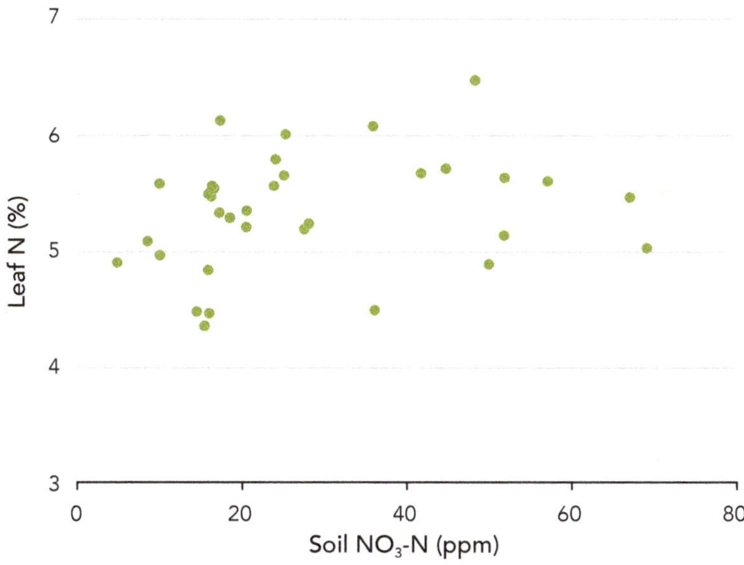

FIGURE 9.2.

Relationship between soil NO₃-N (top 1 foot) and leaf N concentration of lettuce at the early heading growth stage.
Source: Bottoms et al. 2012.

The Nitrate Quick Test

Lab analysis is the most accurate method of determining soil nitrate concentration, and it is the preferred method if analytical results can be returned to the grower in a timely manner. However, a semi-quantitative estimate of soil nitrate concentration can be made using the on-farm "nitrate quick test" procedure. The main advantage of the quick test is that results can be obtained in as little as 1 hour after sampling. This quick test has been effectively used on a wide diversity of mineral soils (below 5 percent organic matter) but has not been calibrated for peat soils.

The materials required for the quick test are volumetrically marked plastic centrifuge tubes (50-milliliter volume), which can be purchased in packs of as few as ten tubes and can be reused indefinitely. The soil extraction solution requires calcium chloride, a chemical that is also used in home canning and brewing. Lastly, nitrate-sensitive test strips, which are sold in packages of 25 to 100 strips, are required to estimate the nitrate concentration. Several brands of test strips are available; any brand should be adequate to the task, provided the strips can read concentrations up to approximately 50 ppm NO_3-N, or 200 ppm NO_3. These items can be purchased online, and vendors can be identified through an internet search.

Quick Test Procedure

1. Prepare a composite soil sample that is representative of the field by collecting and thoroughly blending at least fifteen soil cores from around the field.

2. Make the extracting solution by dissolving approximately 6 grams of calcium chloride (about 1 level teaspoon) in 1 gallon of deionized or distilled water. Precise proportions of calcium chloride and water are not important; the calcium merely acts to flocculate soil particles.

3. Fill a centrifuge tube to the 30-milliliter level with the extracting solution.

4. Slowly add blended soil to the tube until the level of the solution rises to the 40-milliliter level. It is critical that the soil you test is representative of the sample. For moist clay soils that are difficult to blend, pinch off small pieces from each soil core and add to the tube. Testing duplicate samples minimizes variability. Cap tubes tightly and shake vigorously until all soil clods are thoroughly dispersed.

5. Let the sample sit until the soil particles settle out and a zone of clear solution forms at the top of the tube. This may take only a few minutes for sandy soils or an hour or more for clay soils.

6. Dip a nitrate test strip into the clear solution, shake off excess solution, and wait the length of time specified (usually 60 seconds) before reading

the strip. Compare the color that develops on the strip with the color chart provided with the strips. The strip color will continue to darken with additional time, so make the determination quickly.

Interpretation of Results

Converting the strip reading to an estimate of ppm NO_3-N in dry soil requires a correction factor to account for the dilution of the sample with the extracting solution. Since the amount of moisture in the soil sample affects the dilution, an adjustment must be made based on the soil texture and moisture content. Also, take care to determine whether the nitrate test strips are calibrated in ppm NO_3-N or ppm NO_3. The equation to adjust the strip color reading to ppm NO_3-N on a dry soil basis is

strip reading ÷ correction factor (from the table below) = ppm NO_3-N in dry soil.

Correction factors for converting test strip color reading to dry soil NO_3-N

In choosing the correct correction factor, moist soil is close to field capacity, while dry soil is closer to the permanent wilting point. The calculation for a loam soil near field capacity moisture content registering a test strip color of 60 ppm NO_3 is

$$60 ÷ 2.0 = 30 \text{ ppm dry soil } NO_3\text{-N.}$$

A dry clay soil registering a test strip color of 20 ppm NO_3-N would be

$$20 ÷ 0.5 = 40 \text{ ppm dry soil } NO_3\text{-N.}$$

Soil texture	Correction factor for strips reading in NO_3		Correction factor for strips reading in NO_3-N	
	Moist soil	*Dry soil*	*Moist soil*	*Dry soil*
sand	2.3	2.6	0.52	0.59
loam	2.0	2.4	0.45	0.54
clay	1.7	2.2	0.38	0.50

see Lazicki and Geisseler (2017) for more information.

Obtaining a representative soil sample for PSNT requires collecting and blending at least fifteen soil cores to overcome spatial variability within the field. Cores should be collected from different positions within the bed, avoiding areas where fertilizer bands may have been recently applied. Appropriate soil depth depends on the situation. A sample depth of at least 12 inches is usually appropriate; for deep-rooting crops (e.g., tomato, cole crops, cucurbits), sampling the second foot may also provide useful information. When evaluating soil NO_3-N data from the second foot of the soil profile, remember that it may be 40 to 50 days after planting before crop roots can effectively recover N from that soil depth (see fig. 2.4). If irrigation is poorly managed before that time, some of the deep NO_3-N may be leached before the crop can access it.

Residual soil nitrate data can inform N management in several ways. One is to calculate a fertilizer credit for residual soil NO_3-N and adjust the seasonal N fertilization rate accordingly; this approach is best suited to a situation in which PSNT is done only once, early in the season. Analytical results (usually reported in ppm NO_3-N on a dry soil basis) must be converted to pounds of N per acre. The conversion is

ppm soil NO_3-N x sample depth x soil weight = pounds of soil NO_3-N per acre,

where the sample depth is in feet and soil weight is in millions of pounds per acre-foot.

Mineral soils typically weigh from 3.5 to 3.8 million pounds per acre-foot. Therefore, if the top foot of soil contains 20 ppm NO_3-N and the soil is estimated to weigh 3.6 million pounds per acre-foot, there would be approximately 72 pounds of NO_3-N per acre in that top foot of soil.

There is no firm rule as to what fraction of residual soil NO_3-N can safely be credited toward the N fertilizer requirement, but approximately 50 to 70 percent is a reasonable estimate under typical field conditions. It is important to understand the distinction between a fertilizer credit and the fraction of residual soil NO_3-N likely to be taken up by the crop. A fertilizer credit is simply an estimate of the relative availability of residual soil NO_3-N and fertilizer N; field-specific factors govern the actual uptake efficiency from either N source.

A second approach to using PSNT is to compare residual soil nitrate with a threshold value to determine whether N fertilization can safely be delayed or reduced; this approach is best suited for situations in which PSNT sampling is done more than once per season. Extensive California research has documented that sidedress or fertigated N can be delayed in fields with more than 20 ppm residual soil NO_3-N (Bottoms et al. 2012; Breschini and Hartz 2002; Hartz et al. 2000). This threshold is broadly applicable across a range of common vegetable crops because 20 ppm represents enough N to supply crop N uptake requirements for an extended period. Residual soil NO_3-N of 20 ppm in the top foot of soil typically represents from 70 to 80 pounds of N per acre, depending on the soil bulk density. Given the relatively low rate of N uptake during the first half of the growing season (see fig. 2.2), this means that a 20 ppm NO_3-N PSNT value prior to first side-dressing indicates that crop N uptake can be met for at least 2 to 3 weeks from residual soil nitrate. From midseason until harvest, crop N uptake is much faster (typically 3 to 6 pounds of N per acre per day, depending on the crop), but a midseason PSNT value greater than 20 ppm NO_3-N should still be adequate to carry a crop for at least 10 to 14 days. In fields with residual soil NO_3-N below 20 ppm NO_3-N, applying only enough N to bring the soil up to that threshold concentration has proven effective (Breschini and Hartz 2002).

The combined effect of crop N uptake and possible nitrate leaching by irrigation reduces soil NO_3-N over time, and fertilization at some point in the season may be needed even in fields with high initial soil NO_3-N concentration. PSNT sampling can be repeated to predict the need for N application throughout the season. Because summer-grown head lettuce has a relatively short growth period (60 to 70 days) and moderate N uptake requirement (about 120 to 160 pounds of N per acre), sampling twice would usually be adequate to provide season-long guidance. Broccoli or celery, with longer growing seasons and higher N uptake, may require more than two samplings to optimize N management.

The PSNT threshold value of 20 ppm does not mean that crops require that level of soil N availability to grow at peak rates. Vegetable crops can maintain peak growth rates until the soil NO_3-N concentration is depleted to a much lower level. It is not uncommon for soil NO_3-N in coastal leafy greens fields to be drawn down to 5 to 10 ppm before harvest (Bottoms et al. 2012; Smith et al. 2016). This is an important point because if fields are managed to maintain at least 20 ppm NO_3-N right up to harvest, a large amount of soil NO_3-N will be available to be leached by irrigation to establish the next crop or by winter rainfall.

FIGURE 9.3.

A suction lysimeter in a lettuce field. Drawing a vacuum pulls soil solution into the lysimeter. Photo: Richard Smith.

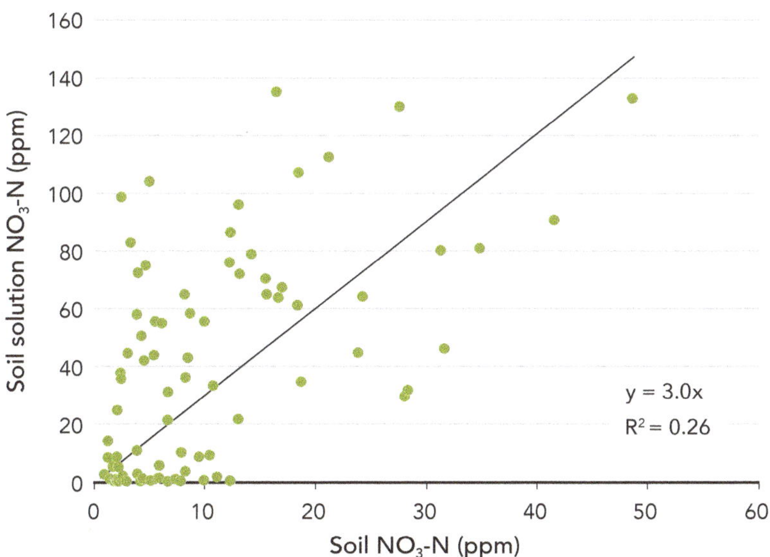

FIGURE 9.4.

Relationship between NO_3-N in soil solution collected with suction lysimeters and root zone soil NO_3-N obtained by traditional soil sampling.

SOIL SOLUTION MONITORING

Another in-season soil monitoring approach is to use a suction lysimeter, also referred to as a soil solution access tube (fig. 9.3). A suction lysimeter is simply a PVC tube with a porous ceramic tip. This instrument is installed in the field with the ceramic tip buried in the active root zone. Periodically a vacuum is drawn in the tube using a hand pump. The vacuum pulls soil solution into the tube from the surrounding soil; this solution may be analyzed for a variety of nutrients, the most common of which are NO_3-N, PO_4-P, and K.

This technique has several significant shortcomings, the most serious of which is a high degree of variability in the measurement. A suction lysimeter draws a sample only from the area immediately adjacent to the ceramic tip; consequently, the sample collected may represent only a couple of inches around the tip. Nutrients are not uniformly distributed throughout the root zone: ions like NO_3^- may be stratified both by the method of fertilizer application (banding versus broadcasting or fertigation) and by irrigation patterns. This spatial variability means that a sample of soil solution drawn from a very small area may not accurately represent the nutrient status of the whole root zone.

Figure 9.4 illustrates this point. The graph shows data from a fertility trial conducted in a drip-irrigated processing tomato field (Hartz 2009). In this trial, weekly soil solution NO_3-N data obtained using suction lysimeters were compared with the soil NO_3-N concentration of composite samples of six root soil cores collected within 30 feet of the lysimeter. Although statistically correlated, the relationship between the soil solution NO_3-N by suction lysimetry and the NO_3-N concentration of composite soil samples was so poor as to make the lysimeter sampling completely unreliable. In theory one might overcome this spatial variability issue by installing a number of lysimeters in each field, but it is impractical to deploy enough lysimeters to produce reliable results.

Another limitation of suction lysimetry is that monitoring soil solution PO_4-P and K provides little information that could not be inferred from a traditional preplant soil test. Since soil solution PO_4-P and K concentration exist in equilibrium with the other P and K pools in the soil (see chapters 5 and 6), soil solution values for PO_4-P and K should not fluctuate wildly over the growing season as NO_3-N can. In fact, one should be able to predict in-season soil solution PO_4-P and K from preplant soil tests with reasonable accuracy.

The bottom line is that while suction lysimetry may seem like a scientific monitoring tool, the reality is that it provides data of questionable accuracy or utility. Adjusting fertilization based on soil solution monitoring often results in wasteful fertilizer application.

REFERENCES

Bottoms, T. G., M. P. Bolda, M. L. Gaskell, and T. K. Hartz. 2013. Determination of strawberry nutrient optimum ranges through DRIS analysis. HortTechnology 23:312–318.

Bottoms, T. G., R. F. Smith, M. D. Cahn, and T. K. Hartz. 2012. Nitrogen requirements and N status determination of lettuce. HortScience 47:1768–1774.

Breschini, S. J., and T. K. Hartz. 2002. Presidedress soil nitrate testing reduces nitrogen fertilizer use and nitrate leaching hazard in lettuce production. HortScience 37:1061–1064.

Hartz, T. K. 2006. Reevaluating tissue analysis as a management tool for lettuce and cauliflower. CDFA FREP final report. CDFA website, www.cdfa.ca.gov/is/ffldrs/frep/pdfs/completedprojects/03-0650Hartz2007.pdf

Hartz, T. K. 2009. Development of practical fertility monitoring tools for drip-irrigated vegetable production. CDFA-FREP project 06-0626 final report. CDFA Website, www.cdfa.ca.gov/is/ffldrs/frep/pdfs/completedprojects/06-0626Hartz.pdf.

Hartz, T. K., W. E. Bendixen, and L. Wierdsma. 2000. Pre-sidedress soil nitrate testing as a nitrogen management tool in irrigated vegetable production. HortScience 35:651–656.

Hartz, T. K., and T. G. Bottoms. 2009. Nitrogen requirement of drip-irrigated processing tomatoes. HortScience 44:1988–1993.

Hartz, T. K., E. M. Miyao, and J. G. Valencia. 1998. DRIS evaluation of the nutritional status of processing tomato. HortScience 33:830–832.

Hartz, T. K., P. R. Johnstone, E. Williams, and R. F. Smith. 2007. Establishing lettuce leaf nutrient optimum ranges through DRIS analysis. HortScience 42:143–146.

Hochmuth, G., D. Maynard, C. Vavrina, E. Hanlon, and E. Simonne. 2012. Plant tissue analysis and interpretation for vegetable crops in Florida. Gainesville: University of Florida Cooperative Extension Publication HS964. IFAS website, edis.ifas.ufl.edu/ep081

Jones, J. B., B. Wolf, and H. A. Mills. 1991. Plant analysis handbook. Athens, GA: Micro-Macro Publishing.

Lazcano, C., J. Wade, W. R. Horwath, and M. Burger. 2015. Soil sampling protocol reliably estimates preplant NO_3 in SDI tomatoes. California Agriculture 69:222–229.

Lazicki, P. A., and. D. Geisseler. 2017. Soil nitrate testing supports nitrogen management in irrigated annual crops. California Agriculture 71(2): 90–95.

Ludwick, A. E., L. C. Bonczkowski, M. H. Buttress, C. J. Hurst, S. E. Petrie, I. L. Phillips, J. J. Smith, and T. A. Tindall. 2010. Western fertilizer handbook, 9th ed. Long Grove, IL: Waveland Press.

Reuter, D. J., and J. B. Robinson. 1997. Plant analysis: An interpretation manual, 2nd ed. Collingwood, Australia: CSIRO Press.

Sanchez, C. A. 1998. Diagnostic tools for efficient N management of vegetables produced in the low desert. CDFA FREP final report. CDFA website, www.cdfa.ca.gov/is/ffldrs/frep/pdfs/completedprojects/95-0222Sanchez.pdf

Smith, R., M. Cahn, T. K. Hartz, P. Love, and B. Farrara. 2016. Nitrogen dynamics of cole crop production: Implications for fertility management and environmental protection. HortScience 51:1586–1591.

Irrigation Effects on Nutrient Management

I rrigation is an integral part of vegetable production, and efficient nutrient management requires coordination of irrigation and fertility practices. Irrigation interacts with soil fertility in a number of important ways. The irrigation wetting pattern controls the extent and density of rooting, which affects the crop's ability to fully exploit the soil nutrient supply. Irrigation can be a source of nutrients and is often a vehicle for fertilizer delivery (fertigation). Nutrient leaching is closely tied to irrigation management. Therefore, efficient nutrient management is impossible without efficient irrigation management.

IRRIGATION FUNDAMENTALS

The irrigation requirement of vegetable crops is driven by weather conditions and stage of crop development. Reference evapotranspiration (ET_o, the amount of water lost through evaporation and transpiration from a well-watered grass crop covering the entire soil surface) can be calculated from environmental variables (temperature, solar radiation, humidity, and wind speed). Real-time ET_o data are available for the major vegetable production regions through the California Irrigation Management and Information System (CIMIS) from the Department of Water Resources website, cimis.water.ca.gov. Historical ET_o values are also available. Table 10.1 lists average daily ET_o by month for representative locations.

ET_o must be multiplied by a factor that accounts for the stage of crop development to estimate the actual crop evapotranspiration (ET_c) that occurs in a field. This factor is called a crop coefficient (K_c). Because crop water use is primarily driven by the absorption of solar radiation by the foliage, K_c can be estimated from the fraction of the ground surface covered by the crop canopy. Figure 10.1 shows the generic relationship between the percentage of canopy cover and K_c for vegetable crops.

Multiplying the daily ET_o by the K_c estimates the daily ET_c. Seasonal ET_c varies widely among crops (table 10.2), based largely on the length of the growing season and the daily ET_o of the production region. Processing tomato requires approximately 120 days under Central Valley summer conditions (daily ET_o

TABLE 10.1.

Historical CIMIS reference evapotranspiration (ET_o) averages, in inches per day

Location	Jan	Feb	Mar	Apr	May	Jun	Jul	Aug	Sep	Oct	Nov	Dec
Bakersfield	.03	.06	.11	.16	.21	.26	.28	.24	.18	.11	.05	.03
El Centro	.09	.13	.18	.26	.33	.37	.37	.31	.28	.20	.11	.06
Five Points	.03	.06	.11	.17	.21	.26	.28	.24	.18	.11	.05	.03
Salinas	.05	.07	.09	.13	.15	.16	.16	.15	.13	.09	.06	.04
Santa Maria	.06	.08	.10	.13	.16	.17	.17	.17	.15	.11	.08	.06
Tracy	.03	.06	.09	.15	.20	.24	.26	.22	.18	.10	.04	.02
Ventura	.07	.09	.10	.13	.15	.16	.18	.16	.14	.11	.08	.06
Woodland	.03	.06	.10	.16	.20	.26	.26	.23	.18	.12	.06	.03

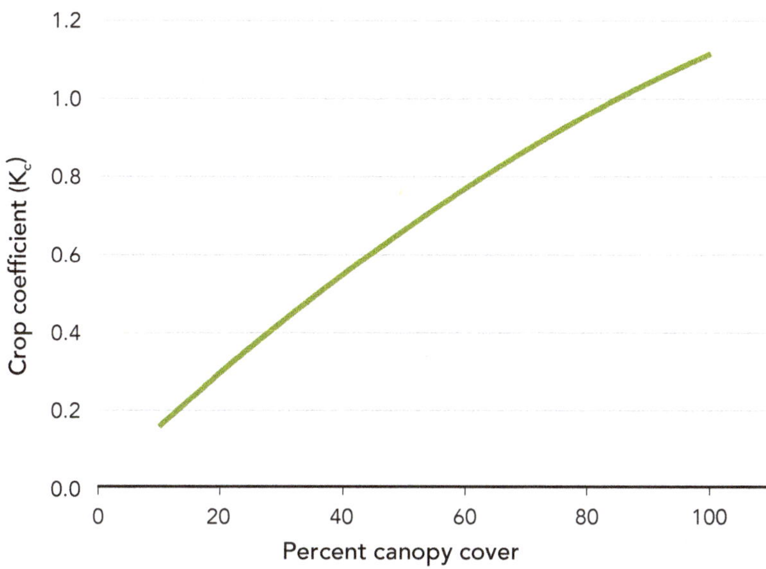

FIGURE 10.1.

Relationship between the crop canopy cover (percentage of ground surface) and the crop coefficient (K_c). The calculation is $K_c = (0.63 + 1.5C - 0.0039C2) \div 100$, where C is the percent canopy cover. Source: Adapted from Johnson et al. 2016.

TABLE 10.2.

Typical ranges of seasonal crop evapotranspiration (ET_c)

Crop	Region	Seasonal ET_c (inches)
broccoli	Central Coast	8–12
cantaloupe	Central Valley	12–16
cauliflower	Central Coast	7–11
celery	Central Coast	7–12
head or romaine lettuce	Central Coast	5–9
processing tomato	Central Valley	22–29

TABLE 10.3.

Effect of soil texture on approximate soil available water-holding capacity and allowable water depletion between irrigations for optimal crop production (in inches/foot)

Soil texture	Available water-holding capacity	Cumulative ET_c allowable between irrigations
sand	0.8	0.2–0.3
sandy loam	1.2	0.3–0.5
loam	1.6	0.4–0.7
clay loam	2.0	0.5–0.8
clay	1.6	0.4–0.7

around 0.25 inches), while summer lettuce requires only about 60 days under coastal conditions (daily ET_o about 0.15 inches). Irrigation method also has an impact: for example, sprinkler irrigation results in greater evaporation from the soil surface than buried drip irrigation.

Due to the inherent inefficiency of field-scale irrigation systems, the amount of irrigation applied must exceed ET_c to ensure that all areas of a field receive adequate water. Field-scale drip systems should be designed to have a distribution uniformity (DU) of 85 to 90 percent, meaning that the driest quarter of the field receives 85 to 90 percent of the field average. The DU of sprinkler and furrow systems is frequently in the 70 to 80 percent range. Dividing the ET_c by the DU estimates the amount of irrigation required to adequately water all portions of the field. For example, a sprinkler-irrigated lettuce field with a seasonal ET_c of 7 inches would require 9.3 inches of irrigation if the D.U. were 75 percent.

Efficient irrigation also requires an irrigation frequency appropriate to field-specific conditions. Vegetable crops are more sensitive to moisture stress than are most other annual crops; allowing more than 25 to 40 percent of available soil moisture to be depleted in the root zone between irrigations can reduce yield potential and in some cases reduce quality. The available water-holding capacity varies by soil texture; table 10.3 lists the approximate available soil water-holding capacities across a range of soil textures and suggests limits on the cumulative soil water depletion allowed between irrigations to minimize crop stress. Rooting depth also influences irrigation frequency. The effective rooting zone of relatively deep-rooted crops (cole crops, cucurbits, tomato, etc.) may be 18 inches or more, while more shallowly rooted crops (lettuce, onion, spinach, etc.) are limited to the top foot.

Drip irrigation frequency can be managed to limit soil moisture depletion to the lower end of these ranges, while with sprinkler or furrow irrigation it is often impractical to irrigate frequently enough to stay within even the upper limit of these ranges. This is a major reason why yield improvements are often observed with drip. However, regardless of the irrigation system used, the larger an individual irrigation, the deeper the wetting front will go. In the process of moving through the soil profile, the wetting front moves nitrate with it. At the same seasonal irrigation volume, using more frequent irrigation tends to minimize the depth to which nitrate will leach and improves crop recovery of applied N.

Depending on the crop grown, the quality of the irrigation water, and the amount of annual rainfall, additional irrigation may be required to maintain soil salinity at a level that allows maximum crop growth. Relative crop salt sensitivities were given in table 3.3. The amount of water required to maintain appropriate salt balance is referred to as the leaching fraction (LF, the proportion of applied water intended to leach salts). In general, if irrigation water salinity is less than 1 decisiemen per meter (dS/m) and annual rainfall is more than 10 inches, salt buildup is a minimal concern. As irrigation water EC increases or annual precipitation decreases, the required LF increases. Under worst-case conditions, a leaching fraction of 20 to 30 percent may be needed to maintain salt balance. However, under most field conditions much less leaching for salt control is necessary, and leaching does not have to be achieved with each irrigation. A more complete discussion of irrigation scheduling is contained in Hanson et al. (2004), while additional information on leaching for salinity control is given in Cahn and Bali (2015).

It should be clear that the potential exists for a substantial volume of leaching to occur under normal field conditions. Although a range of nutrients is contained in leachate, typically only NO_3-N is lost in an agronomically and environmentally significant quantity. Leachate from vegetable root zones commonly contains from 30 to 100 ppm NO_3-N, which is equivalent to approximately 7 to 23 pounds of N per acre-inch of leachate. Careful attention to irrigation system design and maintenance, along with efficient irrigation scheduling, minimizes nitrate loss and reduces N loading to groundwater. Strategically planning leaching events for when soil nitrate concentration is relatively low can minimize nitrate loss.

IRRIGATION EFFECTS ON SOIL NUTRIENT AVAILABILITY

Irrigation management can affect nutrient availability by altering the portion of the soil profile that is wetted. Nutrient availability is stratified in soil, with plant-available P, K, and micronutrients generally decreasing with soil depth. Similarly, the top few inches of soil usually contain more organic matter and are capable of mineralizing more N than soil deeper in the profile. Consequently, the different wetting patterns achieved by irrigation systems can alter nutrient availability. By wetting the entire soil surface, sprinklers maximize the soil volume that roots can exploit and maximize the potential for N mineralization from soil organic matter. Furrow irrigation may be managed to keep the bed tops dry, limiting access to nutrients in that fertile zone. Buried drip irrigation also keeps the surface soil dry while encouraging roots to concentrate close to the drip tape; in this system, nutrient uptake is concentrated in a relatively small portion of the soil profile. By increasing crop yield potential while keeping the surface soil too dry for root activity, drip irrigation may substantially increase the fertilizer requirement.

IRRIGATION WATER AS A SOURCE OF NUTRIENTS

The nitrate concentration in irrigation wells is increasing across California. Many wells now exceed 10 ppm NO_3-N, with some exceeding 40 ppm. In soil, the nitrate added by irrigation water is indistinguishable from that applied as fertilizer or mineralized from soil organic matter. However, there has been widespread reluctance on the part of growers to fully credit irrigation water NO_3-N as the equivalent of fertilizer N. This reluctance stems mainly from the recognition that irrigation efficiency is often poor, and some fraction of irrigation water nitrate is likely to leach before it can be used by the crop. While inefficient irrigation clearly can reduce the fraction of the irrigation water NO_3-N taken up by a crop, the same can be assumed for N applied as mineral fertilizer. In determining a fertilizer credit for irrigation water NO_3-N, the important question is its relative efficiency of uptake by the crop compared to fertilizer N. Cahn et al. (2017) conducted trials in drip-irrigated coastal vegetable fields to answer that question. Treatments where the only N source was irrigation water NO_3-N were compared with standard N fertigation treatments. Across four field trials, irrigation water NO_3-N was at least as efficiently taken up by crops as was fertilizer N. This was true even in treatments that had leaching fractions of above 40 percent.

One reason for such efficient uptake of irrigation water NO_3-N is the fact that it is a "just in time" N application technique. Both the crop irrigation requirement and crop N uptake increase as the season progresses, so the rate of N applied in the water rises in tandem with crop N uptake. Also, N fertigation events may apply enough N to supply crop uptake for several weeks; during that time the fertigated N may be more prone to leaching than NO_3-N applied continuously by irrigation.

There are several ways to calculate a fertilizer credit for irrigation water NO_3-N. A conservative approach would be to credit 100 percent of the N contained in the volume of water transpired by the crop. The logic of this approach is simple: transpiration efficiently brings the NO_3-N content of irrigation water into direct contact with roots. The fertilizer credit would be calculated by multiplying the estimated inches of seasonal ET_c by the nitrate concentration of the irrigation water using these formulas:

ppm irrigation water NO_3-N x 0.227 = lb N/ acre-inch

lb N/acre-inch x inches of seasonal transpiration = lb N/acre of fertilizer credit

The potential agronomic significance of irrigation water N can be illustrated by calculating the fertilizer credit for a broccoli crop grown with well water containing 25 ppm NO_3-N. If the seasonal ET_c is 10 inches (table 10.2), the credit would be

25 ppm NO_3-N x 0.227 x 10 inches = 57 lb N/acre.

For crops like processing tomato that have high seasonal ET_c (> 20 inches), even a relatively low water NO_3-N concentration can represent a significant amount of N.

An alternative approach to calculating a fertilizer credit would be to disregard the NO_3-N in irrigation applied to establish a crop, then credit all of the NO_3-N in water applied after establishment. The logic of this approach is that irrigation for crop establishment is often the least efficient irrigation of the season, and a substantial root system to use the N applied will not be developed until weeks after establishment. Any irrigation water NO_3-N remaining in the root zone after establishment can be accounted for by pre-sidedress soil nitrate sampling (see chapter 9). Once the crop is actively growing, a higher fraction of irrigation water is likely to be transpired, and increasing root depth as the season progresses allows recovery of some NO_3-N from lower soil depths.

When estimating the fertilizer credit of irrigation water, check the units the lab uses to report the nitrate concentration. Labs may report irrigation water concentration as nitrate (NO_3) or as nitrate-nitrogen (NO_3-N). The conversion between these units is

ppm NO_3 ÷ 4.43 = ppm NO_3-N.

Nitrate is usually the only form of N present in irrigation water in an agronomically significant amount, so it is the only N form reported on the typical water test. However, recycled municipal wastewater, which is increasingly being used for irrigation in California, can contain a substantial amount of NH_4-N as well as some organic forms of N that become plant available relatively quickly. Wastewater treatment plants routinely test for these other N sources in addition to NO_3-N, and this information is publically available. One should consider all forms of N when estimating the fertilizer credit for N in recycled water.

FERTIGATION MANAGEMENT

Vegetable growers often inject a portion of the N fertilizer applied to their crops through the irrigation system. This is most often done in drip-irrigated fields, but "water-run" N may be applied in sprinkler or furrow fields as well. Potassium fertigation is also common with drip irrigation. Fertigation has an advantage over other nutrient application methods in that the nutrients can be applied any time during the growing season in close synchrony with crop nutrient uptake. Figure 4.9 shows the mismatch between the timing of N application and crop N uptake in a traditional program of preplant and sidedress N fertilization. Fertigation throughout the crop cycle provides better synchrony with crop N uptake, minimizing potential nitrate leaching loss. While in theory very frequent fertigation (spoon feeding) would be ideal, in practice little added efficiency is gained by fertigating more often than every 7 to 14 days. The exception may be in very sandy soil with limited water-holding capacity, where it is very difficult to control nitrate leaching.

Fertigation has a potential disadvantage in that the uniformity of fertilizer distribution across the field is often lower than if the fertilizer were applied by tractor. The relatively low irrigation DU typically achieved in sprinkler and furrow irrigation makes water-run fertilizer applications particularly inefficient. Even in drip systems, high fertigation DU is not always achieved; poor fertilizer distribution can result from poor irrigation DU or from improper fertigation technique.

When fertigating through a drip system, it is critical that the fertilizer be thoroughly mixed with irrigation water. If fertilizer injection occurs at the field edge as opposed to at the pump, take measures to ensure that fertilizer is well mixed before water reaches a drip line or a branch in the submain. If fertilizer is not adequately mixed before the water flows into the drip lines, some beds may receive more fertilizer than others. Providing extra

distance in the submain for mixing before the first branch in the irrigation system or using injection quills or static mixers can maximize fertilizer DU.

Equally important to achieving high fertilizer DU is allowing sufficient time for all of the injected fertilizer to flush from the drip tape before ending the irrigation set. Irrigators may believe that injecting fertilizer at the end of the irrigation set is the best way to reduce fertilizer leaching. However, because it may take 30 to 60 minutes for fertilizer to travel to the farthest point in an irrigation block, waiting too close to the end of an irrigation set to inject fertilizer can lead to poor DU. The travel time of fertilizer in a drip system can be evaluated by injecting food dye into the system and measuring the time for the dye to arrive at the end of the last drip tape in the irrigation system. Injecting fertilizer during the middle third of a drip irrigation set is a good rule of thumb except for very short (< 3-hour) or very long (> 12-hour) irrigations.

All common N and K fertilizers can be effectively applied through fertigation. Both nitrate and urea are very mobile in soil, so minimizing the amount of water applied after fertigation has stopped is important to prevent deep movement. The high solubility of potassium thiosulfate makes it logistically easier to fertigate, but its relatively high cost may make potassium chloride or potassium sulfate preferable. Fertigating P fertilizer carries the potential risk of clogging drip tape due to precipitation with Ca in the irrigation water. If P fertigation is attempted, using acidic materials (i.e., phosphoric acid) or acidifying the irrigation water before injection reduces precipitation risk. Additional information on effective fertigation is contained in Hanson (2006).

REFERENCES

Cahn, M., and K. Bali. 2015. Managing salts by leaching. Oakland: University of California Agriculture and Natural Resources Publication 8550.

Cahn, M., R. Smith, L. Murphy, and T. Hartz. 2017. Field trials show the fertilizer value of nitrogen in irrigation water. California Agriculture 71:62–67.

Hanson, B. 2006. Fertigation with microirrigation. Oakland: University of California Agriculture and Natural Resources Publication 21620.

Hanson, B., L. Schwankl, and A. Fulton. 2004. Scheduling irrigations: When and how much water to apply. Oakland: University of California Agriculture and Natural Resources Publication 3396.

Johnson, L., M. Cahn, F. Martin, S. Benzen, and B. Farrara. 2016. Evapotranspiration-based irrigation scheduling of head lettuce and broccoli. HortScience 51:935–940.

Organic Fertility Management

Organic vegetable production continues to increase each year, and for some commodities (e.g., baby greens) it now represents a substantial portion of the industry. In theory, the core differences between organic and conventional fertility management are the prohibition of synthetic fertilizer materials and the emphasis on biological management to build the soil's ability to provide crop nutrients. However, organic management does not change the fundamental processes that govern soil nutrient availability.

NITROGEN MANAGEMENT

Unlike conventional production in which mineral N fertilizers can be applied at any time to meet crop N demand, organic methods rely on organic N forms in soil, cover crops, manures, composts, and animal and plant byproducts. Organic forms of N must be mineralized through microbial action to become plant available. Microbial N transformations in soil are complex and relatively slow (see chapter 4). When any organic material is incorporated into soil, N release from that material takes place over weeks or months; the fraction of N mineralized depends on the characteristics of the material and soil temperature and moisture. While educated guesses can be made about these processes, organic management inevitably increases uncertainty about the timing and adequacy of soil N availability. Synchronizing the timing of soil N availability and crop N demand is one of the central challenges of organic production (Gaskell et al. 2006; Mikkelsen and Hartz 2008).

Cover Crops

Cover crops can be a major source of N in organic systems. Most California scenarios involve overwinter cover crops, although in some crop rotations a cover crop may fit at another time of year (Wang et al. 2008). Legume cover crops are widely used for their ability to fix atmospheric N, but nonlegumes (mostly grasses or mustards) are also used for their ability to scavenge mineral N from the soil profile. Mixtures of legumes and grasses are commonly used; since different cover crop species have different environmental tolerances, a cover crop mixture may help ensure good growth over a range of environmental conditions. The amount of N fixed and scavenged by a cover crop varies widely, but in most circumstances a cover crop contains from 100 to 200 pounds of N per acre in its biomass at the time of soil incorporation.

Once the cover crop is incorporated into the soil, N mineralization begins at a rate that is affected by the cover crop C:N ratio. Cover crop residue with a C:N ratio greater than 20 mineralizes slowly and may induce a period of N immobilization just after soil incorporation. Net N mineralization should be immediate and rapid for crop residue with a C:N ratio below 15. Because residue of all common cover crop species has a relatively consistent carbon content (approximately 40 percent of dry weight), knowing just the N concentration allows one to predict the N mineralization behavior. Residue with an N concentration below 2 percent (C:N ratio greater than 20) may induce N immobilization initially, while residue above 3 percent N (C:N ratio less than 15) should mineralize rapidly. As long as adequate soil moisture is present for microbial activity, N mineralization will be most rapid in the first month following incorporation of residue, with the rate of mineralization slowing dramatically thereafter (see fig. 4.4). Legume cover crops generally have higher

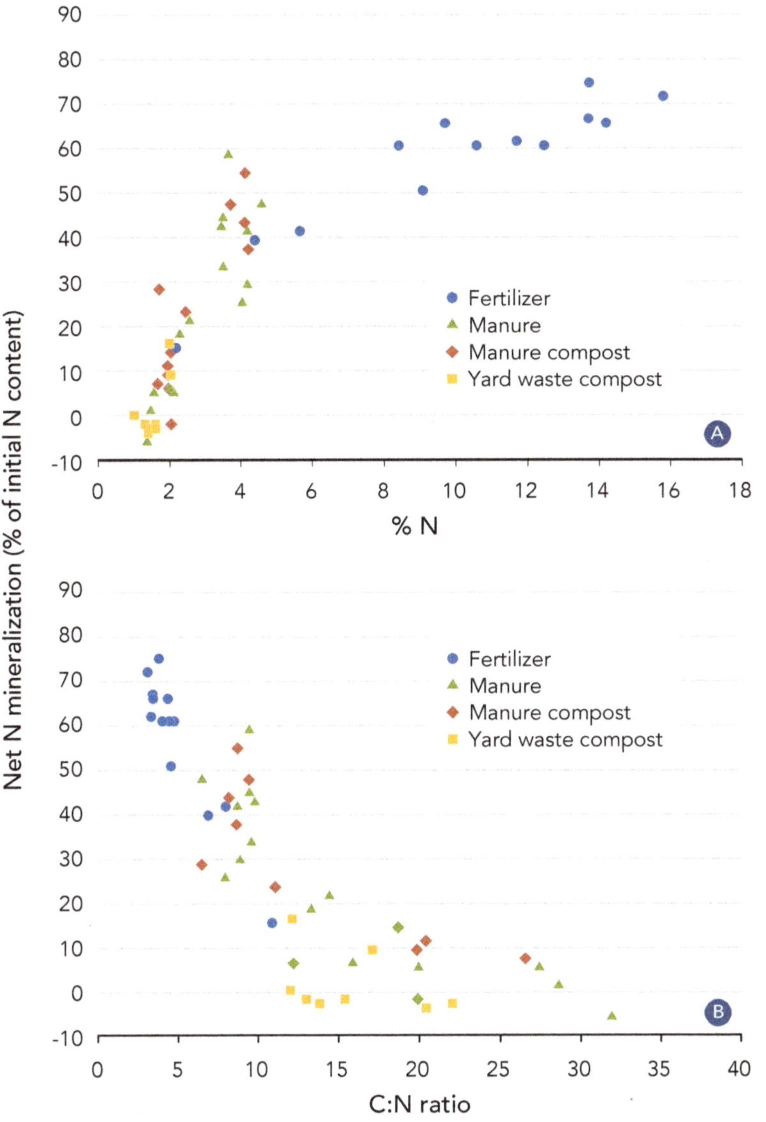

FIGURE 11.1.

Net N mineralization of manures, composts, and organic fertilizers in a lab incubation of 8 to 10 weeks. Results are expressed as a percentage of the initial N content.
Sources: Castellanos and Pratt 1981; Gale et al. 2006; Hartz and John-stone 2006; Hartz unpublished data.

N concentration than nonlegumes, but their mineralization behavior is similar at the same N concentration. The fraction of cover crop N that will be mineralized in the growing season following incorporation typically ranges from 30 to 60 percent (Sullivan and Andrews 2012).

It should be noted that cover crops are less widely used in California organic production than one might expect. In coastal areas, high land values incentivize the production of multiple cash crops per year, which requires tight production schedules; this discourages routine cover cropping. In the Central Valley, organic processing tomato production is seldom preceded by a winter cover crop (Castro and Hartz 2015); again, this is largely due to concern that cover crop residue management may complicate spring planting schedules. As a result, many organic growers rely on the application of organic amendments as the basis for their fertility program.

Manures, Composts, and Organic Fertilizers

A wide variety of manures and composts are used in organic vegetable production. Due to microbial food safety concerns, manures are generally either composted or heat-treated before use. Additionally, many different organic fertilizers are used, most of which contain animal byproducts of some type (fishery waste, feather or blood meal, etc.). Figure 11.1A shows the percentage of N content mineralized by a range of organic materials over the course of a lab incubation of 8 to 10 weeks at 70° to 75°F; these values approximate the N mineralization likely to occur in a field season after soil incorporation. Nitrogen concentration is the best predictor of N mineralization potential. Material containing less than 2 percent N is typically nearly nitrogen neutral, mineralizing or immobilizing no more than 10 percent of its N content. Dairy manure compost and yard waste compost commonly fall in this category. As the N concentration of an organic material increases above 2 percent, the N mineralization potential increases substantially. Organic fertilizers with a high N content (guano, feather meal, fish byproducts, etc., which are typically greater than 8 percent N) mineralize the majority of their N in the growing season after application. There is also a correlation between C:N ratio and N mineralization (fig. 11.1B). Materials with a C:N below 10 are likely to mineralize quite rapidly. Products with a C:N greater than 15 are typically close to nitrogen neutral, mineralizing less than 10 percent of their N content.

The rate at which N is mineralized from organic amendments like those in figure 11.1 is most rapid in the initial weeks after soil incorporation. The majority of the seasonal N contribution of an amendment is likely to be

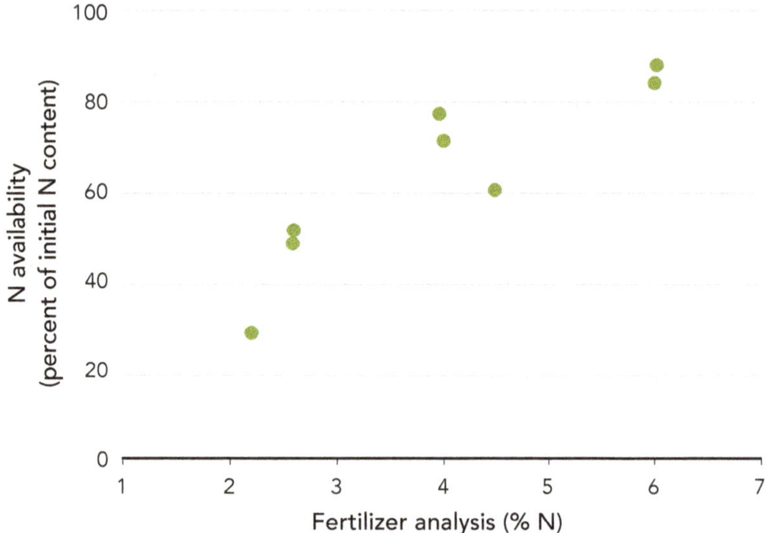

FIGURE 11.2.

N availability from liquid organic fertilizers over a 4-week lab incubation, expressed as a percentage of initial N content.
Sources: Hartz et al. 2010; Hartz unpublished data.

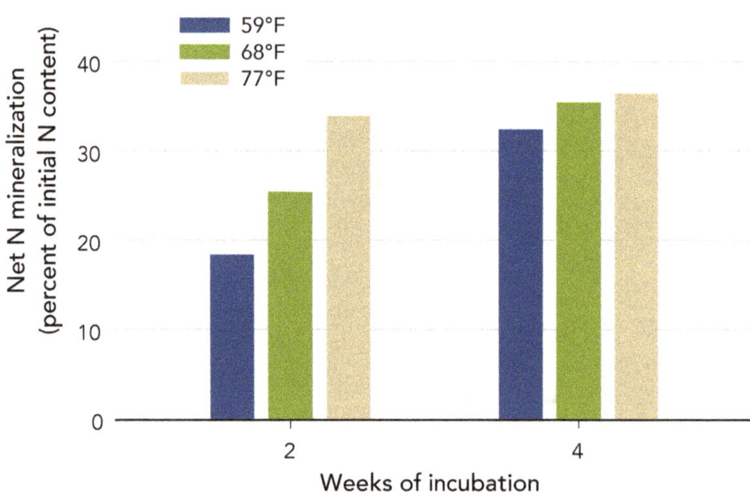

FIGURE 11.3.

Effect of soil temperature on the rate of net N mineralization from broccoli and lettuce crop residue. Bars indicate the mean of the two residues incubated in soil at field capacity moisture content.
Source: Hartz unpublished data.

mineralized within a month after incorporation. Composting, whether actively managed or passively accomplished by stockpiling or aging, tends to reduce N mineralization potential. This is because much of the easily degradable C and N has already been metabolized by microbes during the composting process.

As the use of drip irrigation has increased in organic production, interest in organically approved liquid fertilizers that can be fertigated has also increased. Many liquid organic fertilizers are now available, and they differ from traditional dry organic fertilizers in several important ways. First, liquid organic fertilizers contain a mixture of soluble and particulate N (Hartz et al. 2010). Particulate N, which can exceed 20 percent of the total N content, may pose a risk of plugging emitters if the drip system filtration is inadequate; conversely, effective filtration may remove some particulate N, reducing the N-supplying potential of the material. Second, liquid organic fertilizers are typically made from high-N materials that mineralize rapidly, but since the formulations contain a substantial amount of water, the products' N concentration is deceptively low. Figure 11.2 shows N mineralization from five representative liquid organic fertilizers in a lab incubation. Despite relatively low N analyses, these products mineralized 40 to 90 percent of their N content in just 4 weeks, a rate of mineralization that was much more rapid than dry products of similar N concentration (compare with fig. 11.1A).

Soil Temperature and Moisture Effects

Temperature influences biological activity in soil and therefore has an effect on the speed of N mineralization from crop residues and organic amendments. However, that effect is modest. Figure 11.3 shows the effect of soil temperature on net N mineralization from vegetable crop residues. In this lab experiment, two residues (broccoli and lettuce) were incubated in a representative field soil at constant temperature over 4 weeks. The temperature range used (59° to 77°F) covered the range of spring and summer soil temperatures in the Central Valley and Central Coast production areas (see fig. 4.6). Strong temperature effects on N mineralization were apparent after 2 weeks, but after 4 weeks temperature effects were minor. This result can be explained by the fact that organic material (crop residue, manure, compost, etc.) contains N bound in different types of complex organic molecules, some

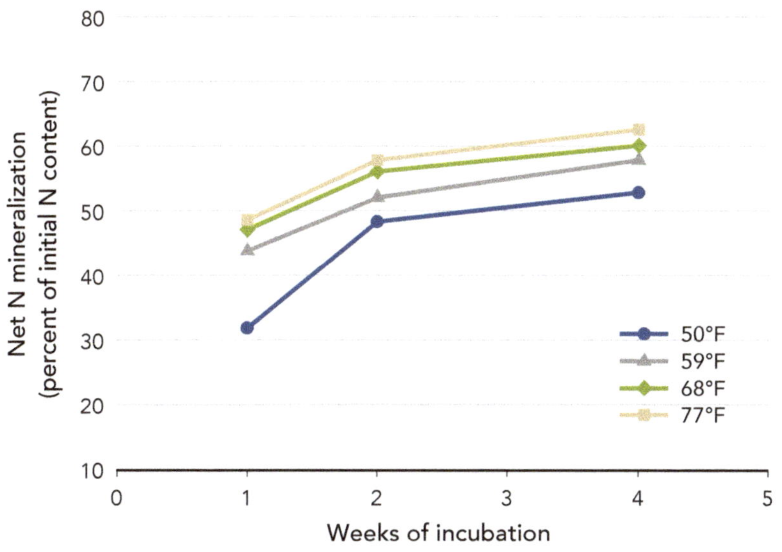

FIGURE 11.4.

Effect of soil temperature on net N mineralization from high-N organic fertilizers. Data are means of four materials (fish powder, blood meal, feather meal, and guano).
Source: Adapted from Hartz and Johnstone 2006.

of which are labile (easily broken down) and others that are more resistant to breakdown. Low soil temperature lengthens the time required for soil microbes to metabolize the labile N, but once all labile N has been mineralized, the rate of additional mineralization is slow regardless of temperature.

Soil temperature has an even smaller effect on the N mineralization behavior of common high-N animal byproducts used as organic fertilizers (fig. 11.4). A substantial fraction of the N in products like fish powder or guano is contained in very simple molecules that can be rapidly broken down by enzymatic hydrolysis. Even at 50°F, most of this highly labile N is mineralized within 2 weeks. These high-N fertilizers are relatively fast acting regardless of soil temperature.

Soil moisture also affects the breakdown of organic materials. Microbial activity is most rapid near field capacity moisture content, so the moist soils maintained in vegetable production support strong microbial activity. However, when crop residue or organic amendments are incorporated into dry soil, minimal breakdown and N mineralization will occur until the soil is rewetted. Conversely, excessive soil moisture can result in a substantial loss of mineral N. Nitrogen loss may be mostly from denitrification in heavy, poorly draining soil or from leaching in lighter soil.

Application technique can also affect N mineralization of organic materials. Soil incorporation is required to maximize N mineralization. While broadcasting organic fertilizers may be convenient, as is sometimes done in leafy greens production, the tradeoff is much slower breakdown and delayed N mineralization.

Nitrogen Mineralization from Soil Organic Matter

In addition to the N mineralized from recently incorporated cover crops and organic amendments, in-season N mineralization from soil organic matter can be significant. Organic production practices tend to increase soil organic matter and soil organic N. Therefore, one would expect organically managed soils to have a higher N mineralization capacity than similar soils managed conventionally. However, California research has shown that the differences in N mineralization potential between conventionally and organically managed soils tend to be modest for several reasons. First, the increase in soil organic matter due to organic production practices is usually relatively small. For example, 10 years of organic management (including cover cropping and compost application) in a Sacramento Valley corn-tomato rotation increased soil organic matter from approximately 1.5 to 1.9 percent in the top 6 inches of soil (Kong et al. 2005); averaged over the top foot of soil, the increase would be smaller. Second, the relative mineralization potential of soil N in organic and conventional soils does not differ very much. In a 4-week incubation experiment with thirty-five Central Valley soils, Castro and Hartz (2016) found that net N mineralization in organic and conventional soils averaged 2.1 and 1.7 percent of soil organic N, respectively. In a field setting, this difference would represent only 20 to 30 pounds more N mineralization per acre in organically managed mineral soils than in similar soils managed conventionally.

The bottom line is that the potential for in-season soil N mineralization, while somewhat higher in organically managed soils, is quite limited compared to the N uptake requirement of vegetable crops. Averaged across a cropping season, the typical organically managed California soil is unlikely to mineralize more than 1 to 2 pounds of N per acre per day from soil organic matter. By comparison, crop N uptake requirement can reach 4 to 6 pounds per acre per day. This means that during the cropping season crop N uptake usually outruns the soil's ability to mineralize N. To avoid N deficiency, either the soil must contain a substantial amount of mineral N at the start of the season (a "bank account" to draw on) or a fast-acting organic fertilizer must be applied to augment soil N mineralization.

In-Season Nitrogen Monitoring

Relatively large amounts of soil mineral N (mostly in nitrate form) may be present in organic fields at the beginning of a cropping season. That mineral N may be a combination of residual soil NO_3-N carried over from the prior cropping season, mineralization from soil organic matter between seasons, and N mineralized from recently incorporated crop residue and organic amendments. As is the case with conventional fields, organic fields differ widely in early season soil NO_3-N content. In a survey of Sacramento Valley organic processing tomato fields, Castro and Hartz (2015) found that soil NO_3-N after transplanting ranged from approximately 30 to 150 pounds of N per acre in the top foot of soil and 40 to 230 pounds of N per acre in the top 2 feet of soil. Not surprisingly, in this study post-transplant soil NO_3-N testing was helpful in predicting which fields were in danger of becoming N deficient later in the season (fig. 11.5). Fields with soil NO_3-N below about 10 ppm were at elevated risk of becoming N deficient, while those above about 15 ppm generally had sufficient N reserves to maximize crop yield. Working with organic broccoli production, Muramoto et al. (2011) also found that early-season soil NO_3-N sampling was useful in predicting crop response to in-season N fertilization. Currently, processing tomato is the only organically grown crop for which a soil nitrate response threshold has been determined. However, organic growers could develop useful information on their own by soil testing fields over several years and correlating crop response with early-season soil NO_3-N.

Early-season soil NO_3-N testing can remove some of the guesswork involved in estimating the N contributions from cover crops or amendments. In most cases at least a month would pass between incorporation of these inputs and post-establishment soil sampling, so the soil sample should capture the rapid initial N mineralization from these materials. If the soil sample is taken within a couple of weeks after planting, a predicted N deficiency could be effectively addressed by sidedressing a high-N fertilizer like feather meal. For long-season crops like processing tomato, N sidedressing as late as 4 to 6 weeks after transplanting can effectively address an N deficiency (Castro and Hartz 2015).

Plant tissue testing has the same limitations in organic culture as in conventional production. Leaf N concentration provides a valid estimate of current plant N status, but a leaf N value within the sufficiency range does not provide actionable information on future N fertilization requirements (see chapter 9). It is true that low N availability is more common in organic production and that an organic grower could encounter low leaf N early enough in the season to adjust N fertilization. However, soil NO_3-N monitoring is the more practical tool to gauge the need for in-season N fertilization because it can be done very early in the season, providing more time for an applied organic fertilizer to mineralize N (see chapter 9). Late-season leaf analysis could assess the adequacy of N management to inform N management practices for future production cycles.

PHOSPHORUS MANAGEMENT

Organic management can improve soil P availability in several ways. Cover crops can scavenge P from deep in the soil profile and move it to the surface when the residue is incorporated. Also, cover cropping and application of organic amendments can stimulate mycorrhizal fungi, which make

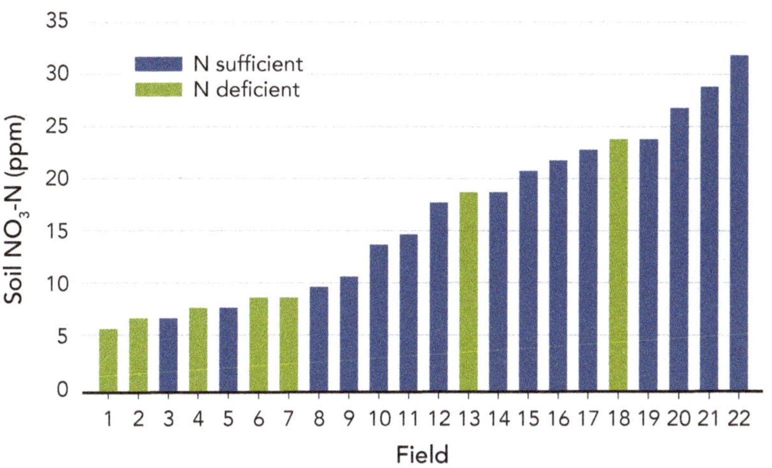

FIGURE 11.5.

Relationship between post-transplanting soil NO₃-N concentration in the top 2 feet of soil and crop N sufficiency at 11 weeks after transplanting. Source: Adapted from Castro and Hartz 2015.

soil P more plant accessible. However, these contributions to soil P availability are typically modest, and P application is often required for peak productivity. Manure, whether fresh or composted, is the primary source of P for most organic growers. The majority (often more than 70 percent) of P in manure is in a mineral form, immediately available for crop uptake (Nelson and Mikkelsen 2008). This means that manure P is nearly as plant available as common inorganic P fertilizers.

Unfortunately, routine application of manure can create an environmental problem. Organic growers use products containing manure both for its N and P content. It is not uncommon for manures or manure composts to have available N:P ratios near 1:1. However, the ratio of N:P uptake by crops typically ranges between about 7:1 to 10:1. This means that reliance on manure to provide substantial N availability can result in excessive P application; over a number of years this can enrich soil P status to the point of posing an environmental hazard to surface water quality. This hazard can be illustrated by considering a survey of organic processing tomato fields in the Sacramento Valley (Castro and Hartz 2015). Most growers relied on manure application, not cover cropping, as the foundation of their N management; at the application rates used, much more P was applied than was removed in harvested fruit. As a result of

this management style over years, 15 of 37 fields had Olsen P above 50 ppm, a level at which P concentration in surface runoff may be environmentally problematic (see chapter 12). Where manure is a significant part of organic fertility management, application rates should be based on crop P requirements, not N content.

Some commonly used organic fertilizers contain bone meal as a source of P. Phosphorus in bone meal is tied up in complex Ca compounds that are highly insoluble in alkaline soil; consequently, bone meal is a highly inefficient source of P in soils with pH above 7.0. Rock phosphate is similarly unsuitable for use in alkaline soil because its P availability is extremely limited.

POTASSIUM MANAGEMENT

Managing K availability in organic culture is straightforward. Potassium exists only in its mineral form and is never bound up in organic molecules. Therefore, K in mineral fertilizers and K in organic materials (crop residue, manure, compost) are equally plant available. Soil test K interpretive standards (see table 3.6) are the same for organic and conventional production. Long-term maintenance of soil K availability is generally less of a problem with organic production because the widespread use of manures and composts (which often have substantial K content) tends to maintain, and in some cases enrich, soil K availability. Unlike N and P fertility, there are no environmental issues associated with K management.

Several mineral fertilizers containing K are approved for organic production, among them potassium chloride (KCl), potassium sulfate (K_2SO_4) and langbeinite (a mixture of K_2SO_4 and $MgSO_4$). Caution is required when using mineral K fertilizers because not all sources of these fertilizers are approved for organic production. Products containing K bound in primary or secondary minerals (e.g., greensand or K powders) must undergo additional weathering to release K. In general, the rate of K release from these sources is too slow to be of agronomic significance (Mikkelsen 2008).

REFERENCES

Castellanos, J. Z., and P. F. Pratt. 1981. Mineralization of manure nitrogen: Correlation with laboratory indexes. Soil Science Society of America Journal 45:354–357.

Castro Bustamante, S., and T. K. Hartz. 2015. Nitrogen management in organic processing tomato production: Nitrogen sufficiency prediction through early season soil and plant monitoring. HortScience 50:1055–1063.

———. 2016. Carbon mineralization and water-extractable organic carbon and nitrogen as predictors of soil health and nitrogen mineralization potential. Communications in Soil Science and Plant Analysis 47 Supplement 1:46–53.

Gale, E. S., D. M. Sullivan, C. G. Cogger, A. I. Bary, D. D. Hemphill, and E. A. Myhre. 2006. Estimating plant-available nitrogen release from manures, composts and specialty products. Journal of Environmental Quality 35:2321–2332.

Gaskell, M., R. Smith, J. Mitchell, S. Koike, C. Fouche, T. Hartz, W. Horwath, and L. Jackson. 2006. Soil fertility management for organic crops. Oakland: University of California Division of Agriculture and Natural Resources Publication 7249.

Hartz, T. K., and P. R. Johnstone. 2006. Nitrogen availability from high-nitrogen-containing organic fertilizers. HortTechnology 16:39–42.

Hartz, T. K., R. Smith, and M. Gaskell. 2010. Nitrogen availability from liquid organic fertilizers. HortTechnology 20:169–172.

Kong, A. Y. Y., J. Six, D. C. Bryant, R. F. Denison, and C. van Kessel. 2005. The relationship between carbon input, aggregation, and soil organic carbon stabilization in sustainable cropping systems. Soil Science Society of America Journal 69:1078–1085.

Mikkelsen, R. 2008. Managing potassium for organic crop production. Better Crops 92(2): 26–29.

Mikkelsen, R., and T. K. Hartz. 2008. Nitrogen sources for organic production. Better Crops 92(4): 16–19.

Muramoto, J., R. F. Smith, C. Shennan, K. M. Klonsky, J. Leap, M. S. Ruiz, and S. R. Gliessman. 2011. Nitrogen contribution of legume/cereal mixed cover crops and organic fertilizers to an organic broccoli crop. HortScience 46:1154–1162.

Nelson, N., and R. Mikkelsen. 2008. Meeting the phosphorus requirement on organic farms. Better Crops 92(1): 12–14.

Sullivan, D. M., and N. D. Andrews. 2012. Estimating plant-available nitrogen release from cover crops. Corvallis: Pacific Northwest Extension Publication PNW 636.

Wang, G., M. Ngouajio, M. E. McGiffen, and C. M. Hutchinson. 2008. Summer cover crop and in-season management system affect growth and yield of lettuce and cantaloupe. HortScience 43:1398–1403.

Nutrient Management and Environmental Protection

Nitrogen and phosphorus loss from fields can represent a serious threat to the environment. In a number of locations in California, surface waters have been adversely impacted as these nutrients leave agricultural fields in surface runoff or tile drainage. Both N and P can stimulate the growth of algae, causing aesthetic issues as well as serious disruptions to aquatic ecosystems. Even low concentrations of these elements in surface water can be environmentally damaging. Waterways as ecologically diverse as the Sacramento–San Joaquin Delta, Elkhorn Slough in Monterey County, and the Salton Sea in Southern California are under regulatory action to reduce nutrient loading from agriculture.

Nitrogen also presents a hazard to groundwater quality throughout the state. Nitrate moves freely with water and is easily leached from the crop root zone with rainfall or irrigation. Groundwater is widely used for drinking water, and the federal government has established a limit of 10 ppm NO_3-N (equivalent to 45 ppm NO_3^-) for human safety. Studies have linked excessive nitrate ingestion to various medical disorders, including methemoglobinemia, in which nitrate interferes with the ability of hemoglobin to carry oxygen throughout the body. Although nitrate pollution can come from industrial and urban sources, a recent UC Davis study identified agriculture as the largest source of N loading to groundwater in the Salinas Valley and Tulare basin (Harter et al. 2012); agriculture is likely to be a significant contributor in other areas of intensive farming as well.

Lastly, gaseous nitrogen lost from vegetable fields to the atmosphere can be environmentally damaging in several ways. Ammonia lost through volatilization can be deposited miles away in natural areas, where it can impact the ecology of lakes and streams. It can also react with other pollutants in the atmosphere to form smog and hazardous particulates. Nitrous oxide (N_2O) release from denitrification represents a significant portion of greenhouse gas impacts from agriculture.

A maze of federal and state laws regulates agricultural contributions to both air and water pollution. However, regulatory pressure currently centers on environmental water quality protection, which is the focus of this chapter.

SURFACE WATER POLLUTION

Nitrogen and phosphorus can be "biostimulatory" in surface water; movement of these nutrients from farm fields into surface waters contributes to eutrophication (stimulation of the growth of algae and other aquatic plants to the detriment of the aquatic ecosystem). Excessive nutrient availability drives cycles of algae growth and dieback, which result in a host of negative effects (production of environmental toxins, fish kills, etc.). To stimulate algae and other aquatic plants, N must be in a reactive form; organically bound N does not contribute substantially to this problem. NO_3-N represents the overwhelming majority of reactive N in agricultural runoff. Current environmental targets for surface water are based on the federal drinking water standard of 10 ppm NO_3-N; however, lower concentrations may be needed to prevent algae blooms in some water bodies. Phosphorus is bio-stimulatory at even lower concentrations; to minimize algae growth, surface water should be less than 0.1 ppm PO_4-P.

Phosphorus-rich sediment deposited in streams or lakes can release enough reactive P over time to contribute to eutrophication.

Soil P availability is the primary driver of soluble P concentration in surface runoff. In an experiment in which runoff was collected from pots of twenty-five California soils exposed to sprinkler irrigation, the runoff PO_4-P concentration was strongly correlated with the soil test P level (fig. 12.1). In this study, runoff PO_4-P was low until soil exceeded approximately 50 ppm Olsen P, with runoff PO_4-P increasing rapidly at higher Olsen P levels. These data suggest that runoff P hazard can be minimized by managing P fertility to avoid excessive soil P accumulation.

Nitrate concentration in runoff is a function of the NO_3-N concentration of the top few inches of soil and the NO_3-N concentration of the irrigation water. Fields receiving heavy N fertilization are more likely to have elevated surface soil NO_3-N than fields receiving low to moderate N inputs. There is a strong positive correlation between irrigation water NO_3-N and NO_3-N in runoff, so special care must be taken to control runoff when using water with elevated NO_3-N concentration. Similarly, it is important to prevent runoff during fertigation, because the N concentration of the applied water can be very high while fertilizer is being injected.

For example, delivering 10 gallons of CAN-17 in 0.2 acre-inches of sprinkler irrigation results in a NO_3-N concentration above 300 ppm.

Fields with tile drainage systems present a particularly challenging problem. Even in well-managed fields, water percolating through the fertilized root zone will inevitably contain NO_3-N far in excess of the 10 ppm drinking water standard; monitoring has shown that tile drainage from vegetable fields commonly contains more than 50 ppm NO_3-N and can exceed 150 ppm (Hartz et al. 2017; Los Huertos et al. 2001). High P concentration would not be expected in tile drainage because most soils can tie up soluble P as it percolates through the profile. However, in fields that have been heavily fertilized over many years, this ability to sequester P can eventually be exhausted, and problematic concentrations of PO_4-P may be found in tile drainage.

Conservation practices may be useful in reducing runoff nutrient losses. Tailwater basins and return systems, vegetated filter strips or drainage ditches, and the use of polyacrylamide (PAM) can be effective in the right situation (Long et al. 2010). However, these practices tend to be more effective in reducing sediment-bound P than in reducing NO_3-N or PO_4-P in runoff. Eliminating runoff is obviously the most effective method of surface water protection.

The high salt content of tile drainage (typically 2 to 5 dS/m or more) limits a grower's ability to recycle that water, so most tile drainage must be released. While biological treatment to remove NO_3-N from tile drainage using constructed wetlands (Osiadacz et al. 2010) or denitrification bioreactors (Hartz et al. 2017) has been demonstrated, the feasibility of farm-scale use of either technology has not been proven. The most practical way for a grower to limit NO_3-N losses in tile drainage is to maximize crop N use efficiency, which usually requires a reduction in N fertilization rates combined with more efficient irrigation. Treatment of tile drainage with alum (aluminum sulfate) can precipitate a substantial percentage of the soluble P content (Bottoms 2013).

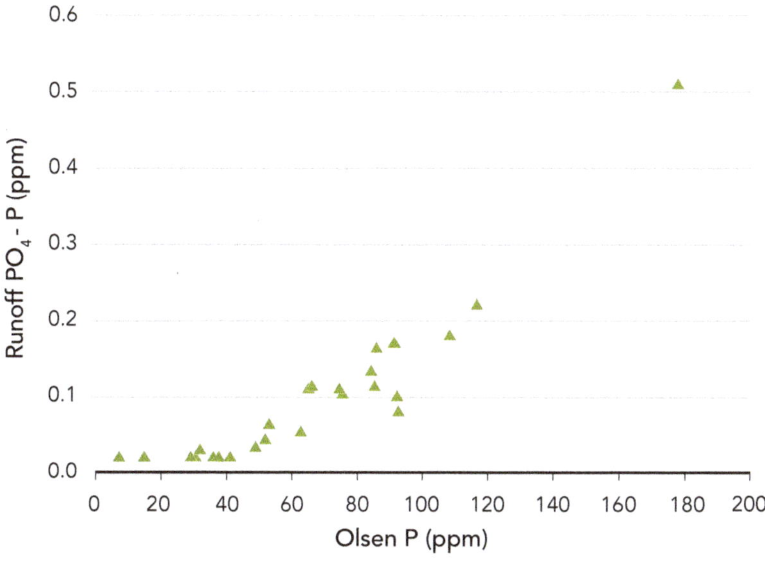

FIGURE 12.1.

Relationship between Olsen P in soil and PO₄-P in surface runoff. Source: Adapted from Hartz and Johnstone 2006.

GROUNDWATER POLLUTION

Nitrate leaching from vegetable fields is a threat to groundwater in many areas of California. The federal drinking water standard of 10 ppm NO_3-N is the regulatory target for groundwater statewide. Nitrate leaching requires both the presence of NO_3-N in the soil and excess water (whether from irrigation or rain) to move it toward groundwater. Effective control of NO_3-N leaching requires attention to both factors.

Fertilized root zones in vegetable fields often contain 10 to 30 ppm NO_3-N on a dry soil basis. Since NO_3^- ions do not adhere to soil particles, all of them are found in the soil solution. In a medium-textured mineral soil at field capacity, the soil solution weighs only about 25 percent as much as the dry soil. This means that the NO_3-N concentration in the soil solution is about four times higher than if expressed on the basis of soil dry weight. Therefore, a silt loam soil with 20 ppm NO_3-N on a dry soil basis would have a soil solution concentration of approximately 80 ppm. Even when diluted by irrigation water, it is clear that leachate from this soil would be far above the 10 ppm NO_3-N environmental standard.

Efficient irrigation management can reduce but not eliminate nitrate leaching. Due to irrigation system nonuniformity and the need to control salinity, nitrate leaching can never be completely prevented. However, the volume of leachate can be minimized by optimizing irrigation design and management. Improving irrigation efficiency may not reduce leachate NO_3-N concentration; in fact, it may increase it. However, the total N loading to groundwater (pounds of NO_3-N per acre) will likely be reduced.

NITROGEN BALANCE

While the environmental standard for groundwater NO_3-N is expressed as a concentration (10 ppm), the most practical way to evaluate trends in water quality on a watershed scale is by estimating the amount of N loading that is occurring. That is why regional water quality control boards are focusing their regulatory programs on a nitrogen balance approach. A nitrogen balance seeks to compare N application to fields (from all sources, including fertilizer, organic amendments, and irrigation water) with the N removed from the field in harvested products. The underlying assumption is that N applied to a field but not removed with the harvest is at risk of eventually moving into the environment, where it can cause damage.

While this assumption is not strictly true, it is not far off the mark. Figure 12.2 illustrates the main features of an agricultural N balance. The relative amounts of N lost by volatilization, denitrification, surface runoff, and leaching vary based on field-specific conditions. Collectively, these four loss pathways usually constitute the vast majority of the difference between N input and harvest N removal, with nitrate leaching usually the largest component of N loss from vegetable fields. The wild card in an N balance calculation is storage of N in the soil. A considerable amount of NO_3-N may carry over in the root zone from one crop to the next. However, across a rotation of several crops there is seldom a net increase in root zone NO_3-N; over time it is either taken up by crops or leached by rainfall or irrigation. Nitrogen storage in

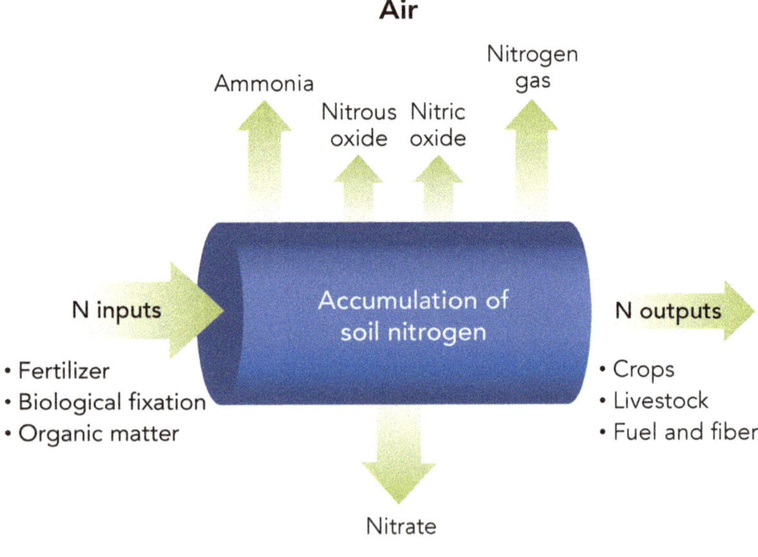

FIGURE 12.2.

Nitrogen balance in agricultural production.
Source: The Fertilizer Institute, tfi.org.

TABLE 12.1.

Simplified N balance of select vegetable crops, based on commercial field monitoring

| Crop | Fertilizer application | Crop uptake | Harvest removal | Difference between N application and | | Reference |
				Crop uptake	Harvest removal	
	N (lb/ac)					
broccoli*	190	330	100	-140	90	Smith et al. 2016a
cauliflower*	230	330	70	-100	160	Smith et al. 2016a
lettuce*	140	130	70	10	70	Bottoms et al. 2012
spinach*	180	120	90	60	90	Smith et al. 2016b
processing tomato	190	270	140	-80	50	Hartz and Bottoms 2009; Lazcano et al. 2015

*Summer coastal conditions.

soil organic matter is a more complex issue. Where conservation tillage, cover cropping, or organic amendments are used, there can be a net storage of N in soil organic matter over time (Kong et al. 2005; Mitchell et al. 2017). However, under conventional vegetable production practices, the amount of N in soil organic matter tends to either remain stable or even decline over time.

While a simple N balance calculation does not account for all of the complexity of the nitrogen cycle, it can be useful in ranking the relative efficiency with which various crops are grown and can provide a comparison of how well different growers manage N inputs. Nitrogen balance can be evaluated either as the difference between N input and crop N uptake or as the difference between N input and N removal in harvested products. Table 12.1 compares the typical N balances of some important vegetable crops. The picture looks reasonably positive if the comparison is between N application and N uptake, with some crops showing a large negative balance (more N uptake than N application). However, from an environmental standpoint the important comparison is N application versus harvest removal. All crops have a large positive balance, meaning that there is a considerable amount of applied N unaccounted for at the end of the season. It is important to consider that table 12.1 includes only fertilizer as an N input; adding the contribution of irrigation water NO_3-N and organic amendments further unbalances the "balance sheet."

Grower-reported N usage shows large differences in N application rates (fig. 12.3). While differences in yields, soil conditions, irrigation practices, and rotational effects may explain some variation in grower fertilizer rates, such large differences in N application clearly suggest that some growers are much more efficient with their N inputs than others. Regulators are likely to focus their attention on those growers whose N application rates exceed industry norms.

Figure 12.3 also demonstrates the substantial role that irrigation water NO_3-N can play in N fertility. Many wells in the Central

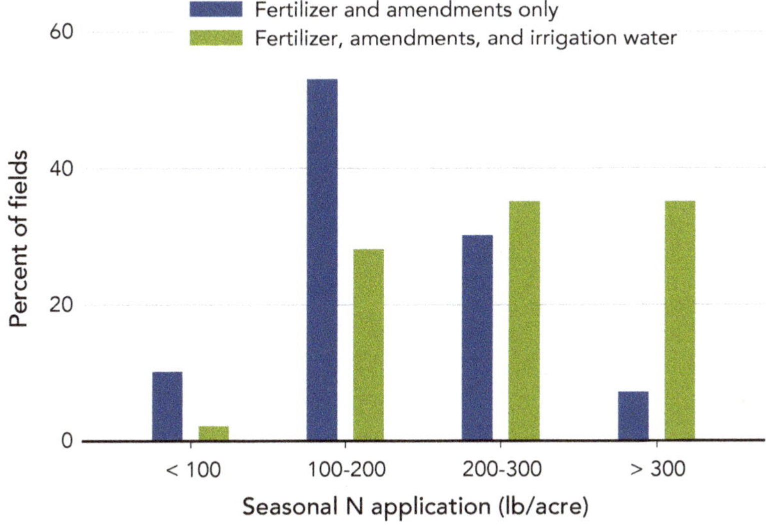

FIGURE 12.3.

Grower-reported N application on lettuce, 2015.
Source: Nitrogen use reports to the Central Coast Regional Water Quality Control Board.

Coast region have elevated NO$_3$-N content; in 2015 the grower-reported irrigation well nitrate content averaged nearly 5 pounds of N per acre-inch, with some wells more than double that. This is not exclusively a coastal issue. Central Valley wells generally have lower nitrate concentrations, but individual wells can be very high, particularly on the east side of the valley. Also, seasonal irrigation amounts are generally higher for vegetable crops produced in the Central Valley than in coastal districts. For example, processing tomato typically receives more than 20 inches of irrigation, whereas less than 10 inches is usually adequate for coastal lettuce. Even wells with modest NO$_3$-N concentration can contribute substantially to overall N supply when seasonal irrigation volume is high.

PRACTICAL STEPS TOWARD IMPROVED NUTRIENT EFFICIENCY

It is clear that environmental regulations will increasingly affect how vegetable crops are fertilized and irrigated. Regulatory activity regarding surface water protection will undoubtedly continue to advance, but for the foreseeable future the regulatory focus will be on reducing nitrate loading to groundwater. The reality is that no meaningful reduction in N loading to groundwater will be possible without a substantial reduction in N fertilizer application. Maintaining peak crop productivity with reduced N fertilization rates will require a more strategic approach to N management, in which N fertilization decisions are made on a field-specific basis rather than on the basis of a standard fertilization program. The following sections detail the most important elements of an efficient N management program.

Realistically Estimate the Crop Nitrogen Requirement

The first step toward efficient N management is to have a realistic estimate of the crop N requirement. Table 2.1 lists ranges of crop N uptake observed in successful commercial fields. Given the fact that these observed N uptake rates undoubtedly include some luxury consumption, using the midpoint of the range should be a safe estimate of the actual crop N uptake requirement for the typical field. Some adjustment for higher-than-normal yield potential based on high plant population (vegetative crops) or prior yield history (fruiting crops) may be warranted. The more complicated question is what fraction of the crop N uptake requirement must come from applied fertilizer. Some generalizations about N fertilizer requirement can be made: for example, the fertilizer requirement may be higher following a wet winter due to the likelihood that residual soil nitrate will be low. However, field monitoring is required for maximum efficiency.

Adjust the Nitrogen Fertilization Rate for Soil Nitrogen Contributions

Nitrogen efficiency requires consideration of potential soil contributions toward crop N requirements. Soil contributions include both the amount of residual NO$_3$-N present at the start of the season and N mineralization of soil organic N during the season. Residual soil NO$_3$-N is by far the most important variable to integrate into a field-specific N management plan; chapter 9 contains an extended discussion on how this can be done. In-season soil N mineralization is a relatively minor factor in soil with low organic matter, but for soils with more than 2 percent organic matter adjustment of N fertilizer rates may be warranted (see the discussion in chapter 4).

Account for Irrigation Water Nitrate-Nitrogen

Irrigation water NO$_3$-N content varies from virtually none to more than 10 pounds of N per acre-inch; in some fields the N input with irrigation can exceed 100 pounds per acre. Since Cahn et al. (2017) documented that irrigation water NO$_3$-N is used by crops as effectively as fertilizer N, adjustment for this input is mandatory to achieve N efficiency. Chapter 10 outlines how to estimate a fertilizer credit for irrigation water NO$_3$-N.

Manage Irrigation Efficiently

As described in chapter 10, each acre-inch of leachate from a fertilized vegetable root zone may carry 10 pounds of NO_3-N or more. Efficient N management is impossible unless irrigation is controlled to minimize in-season leaching.

For most vegetable crops and under most field conditions, it should be possible to produce a good crop with a seasonal N fertilization rate at or in some cases far below the crop N uptake requirement (see table 12.1). However, even where N fertilization is efficiently managed, it is impossible to consistently limit N fertilization to the rate of harvest N removal; there will always be some nitrogen loss to the environment. The key is to keep that loss as small as possible.

REFERENCES

Bottoms, T. G. 2013. Nitrogen management and water quality protection in California lettuce and strawberry production. PhD. diss., University of California, Davis.

Bottoms, T. G., R. F. Smith, M. D. Cahn, and T. K. Hartz. 2012. Nitrogen requirements and N status determination of lettuce. HortScience 47:1768–1774.

Cahn, M., R. Smith, L. Murphy, and T. Hartz. 2017. Field trials show the fertilizer value of nitrogen in irrigation water. California Agriculture 71:62–67.

Harter, T., J. R. Lund, J. Darby, G. E. Fogg, R. Howitt, K. K. Jessoe, G. S. Pettygrove, J. F. Quinn, and J. H. Viers. 2012. Addressing nitrate in California's drinking water with a focus on Tulare Lake Basin and Salinas Valley groundwater. Report for the State Water Resources Control Board to the Legislature. Davis, CA: University of California Center for Watershed Sciences, groundwaternitrate. ucdavis.edu/files/138956.pdf

Hartz, T. K., and T. G. Bottoms. 2009. Nitrogen requirement of drip-irrigated processing tomatoes. HortScience 44:1988–1993.

Hartz, T. K., and P. R. Johnstone. 2006. Relationship between soil phosphorus availability and phosphorus loss potential in runoff and drainage. Communications in Soil Science and Plant Analysis 37:1525–1536.

Hartz, T., R. Smith, M. Cahn, T. Bottoms, L. Tourte, S. Castro Bustamante, K. S. Johnson, and L. J. Coletti. 2017. Wood chip denitrification bioreactors can reduce nitrate pollution in tile drainage. California Agriculture 71:41–47.

Kong, A. Y. Y., J. Six, D. C. Bryant, R. F. Denison, and C. van Kessel. 2005. The relationship between carbon input, aggregation, and soil organic carbon stabilization in sustainable cropping systems. Soil Science Society of America Journal 69:1078–1085.

Lazcano, C., J. Wade, W. R. Horwath, and M. Burger. 2015. Soil sampling protocol reliably estimates preplant NO_3- in SDI tomatoes. California Agriculture 69:222–229.

Long, R. F., B. R. Hanson, A. E. Fulton, and D. P. Weston. 2010. Mitigation techniques reduce sediment in runoff from furrow-irrigated cropland. California Agriculture 64:135–140.

Los Huertos, M., L. E. Gentry, and C. Shennan. 2001. Land use and stream nitrogen concentrations in agricultural watersheds along the central coast of California. Scientific World Journal 1:615–622.

Mitchell, J. P., A. Shrestha, K. Mathesius, K. M. Scow, R. J. Southard, R. L. Haney, R. Schmidt, D. S. Munk, and W. R. Horwath. 2017. Cover cropping and no-tillage improve soil health in an arid irrigated cropping system in California's San Joaquin Valley, USA. Soil and Tillage Research 165:325–335.

Osiadacz, M., W. Brandt, S. Clifton, C. Nicol, D. Nishijima, E. Paddock, V. Pristel, and M. Los Huertos. 2010. Successful treatment systems for removing nitrate concentrations from agricultural runoff. CSU Monterey Bay Watershed Institute Publication WI-2010-07.

Smith, R., M. Cahn, T. K. Hartz, P. Love, and B. Farrara. 2016a. Nitrogen dynamics of cole crop production: Implications for fertility management and environmental protection. HortScience 51:1586–1591.

Smith, R., M. Cahn, and T. K. Hartz. 2016b. Evaluation of N uptake and water use of leafy greens grown in high-density 80-inch bed plantings, and demonstration of best management practices. CDFA Fertilizer Research and Education Program Final Report 12-0362-SA. CDFA website, www.cdfa.ca. gov/is/ffldrs/frep/pdfs/completedprojects/12-0362-SA_Smith.pdf

Index

CPSIA information can be obtained
at www.ICGtesting.com
Printed in the USA
JSHW030038170520
5723JS00005B/12